THE PICK AND
THE PEN

. . . the urge to preach, to criticize, is the essential of successful editorship; it is, by and large, characteristic of all men of active mind . . . T.A. Rickard

A. J. WILSON

MINING JOURNAL BOOKS LIMITED
LONDON

First published 1979
© 1979 MINING JOURNAL BOOKS LTD

ISBN 0 900117 16 8

Text set in 11/13pt Times New Roman, printed in Great Britain by Unwin Brothers Limited, The Gresham Press, Old Woking, Surrey.
A member of the Staples Printing Group.

Contents

Foreword

A history of mining journalism, like the industry it mirrors, is a many-sided story. In this media age, the mining industry depends on the daily press for prompt, accurate and dispassionate reporting of the events which make markets; on the periodical journals for selecting from the mass of available materials those events, technical innovations and ideas which may be of more than passing significance; and on the technical journals for data and insights into the ever-challenging science of winning resources from the earth.

In the 16th century, Georgius Agricola, a physician in Saxony, carefully described the mining practices and procedures he observed in his native land. "Those things we see with our own eyes and understand by means of our senses are more clearly to be demonstrated than if learned by reasoning," he wrote. He felt compelled to explain his differences with the teachings of the philosophers and the mystical works of the alchemists. And he wrote in Latin so that this work might obtain the widest circulation. Copies of his treatise, *De Re Metallica,* we are told, found their way to mining and university centres throughout the known world, including the Americas where the Spanish were already mining gold and silver in the high Andes and in what is now Mexico.

Arthur Wilson has written the story of the mining literature of the years between and of the writers, engineers and editors who continued to observe and report, prod and inform, and through it all to promote and encourage the free and open exchange of technical information which characterizes the mining industry today. To all of these men and women who contributed their talents and their insights toward the development of the mining industry as we know it today, we owe a warm embrace.

The author notes, somewhat wistfully, in a retrospective paragraph that the crusading journalists of the mining press of a prior generation have now almost entirely disappeared. Perhaps so, for the mining periodicals must now concern themselves with the specialized needs of a mature, international industry. But the crusading journalists have not themselves disappeared. Their columns appear in the general media as they write of mine safety, occupational health and the effects of mining on the environment. Few minerals are transparent, but miners still live in glass houses.

Charles F. Barber,   April, 1979.
Chairman,   120 Broadway,
Asarco Inc.   New York, N.Y. 10005.

Preface

This book has been written for three main reasons.

First, as far as I am aware, the subject has never before been dealt with in depth or in detail on an international basis. That, in itself, was a challenge.

Secondly, having been for some 25 years in a middle position between the mining industry and those sections of the world's Press directly concerned with its affairs, I felt that there was a lack of knowledge of the contributions that the media has made, and is continuing to make, to the industry, other than by providing a mirror to its development.

Thirdly, I was concerned that among those countries of the Third World in which mining development is becoming increasingly important, there persists a suspicion that the international Press is a tool of Western capitalism and should be mistrusted or, at least, treated with great caution. By recounting some of the major services which the media has rendered to the industry over many years, I hope that a better understanding and closer relationship can be encouraged to develop between those who control the industry in these countries and the Press. The industry will certainly benefit if this can be achieved.

No book which attempts to cover its subject internationally can be completely comprehensive. There are bound to be some unavoidable omissions. In this book it has, for example, been impossible to probe the development of mining journalism behind the Iron Curtain. In other parts of the world, records have been lost or destroyed during change of ownership or control. Some publishers have found little time to preserve their archives; others have complete records going back nearly 200 years. Some present-day editors who were approached for information were surprised when their researches revealed history of which they had hitherto been unaware.

To all who have helped in assembling material I extend my sincere thanks. In preparing the script I have also drawn liberally on the written works of others: a bibliography is included on page 285 and if there are any omissions I offer my sincere apologies.

It is not practical to name everyone who has helped, but I should like to express special thanks to the following:

The Institution of Mining and Metallurgy for giving me free range of their magnificent London library, which shelves more than 30,000 books on mining, and for the ever-ready assistance of Jennifer Edkins and Mike McGarr of the library staff;

Simon D. Strauss, vice-chairman of Asarco Inc., New York; Si Wakesberg, commodities vice-president of the National Association of Recycling Industries Inc., New York; and Alvin W. Knoerr, of the US Bureau of Mines, Washington DC; for their contributions to Chapter 18;

Mary Anne Niarchos, of *Engineering and Mining Journal*, New York, for her thorough researches into the history of the Journal on my behalf; and to Stanley H. Dayton, the editor, for making such researches possible;

George O. Argall jnr., editor of *World Mining*, San Francisco, for his help in identifying some lesser-known publications of the American West, and more particularly for introducing me to Genevieve Moore (Chapter 15), one of the unforgettable characters of mining journalism;

John M. Dew, of Conzinc Rio Tinto of Australia, and former editor of *Australian Mining*, without whom it would have been difficult to make a start on researches in Australia;

Joe Colgan, of the *Queensland Government Mining Journal*, for his help with the early history of the paper—and for his promise personally to prepare for me some prime Queensland barbecued steak when copies of this book reach Australia;

William Schabas, until recently associate editor of *Canadian Mining Journal*, for his kindness in giving me a preview of the Journal's centennial series in 1978, and for much material relating to B.T.A. Bell (Chapters 6 and 7); and

John Ruth and Dorothy Fabian for help with the early history of metal marketing journalism in the United States.

My thanks are also due to David Goodman for his translation of some early German mining literature.

Finally, I should like to thank friends and former colleagues, both in the mining companies and among the Press in many countries, for indulging my enthusiasm to set down for the record some of the contributions that have been made by the media to the development of the industry, and thereby going some way towards filling a vacuum which I felt existed in its history.

A.J.W.   London, April 1979

CHAPTER ONE

Father
of
Mineralogy

"Many persons hold the opinion that the metal industries are fortuitous and the occupation is one of solid toil, and altogether a business requiring not so much skill as labour ... but as for myself, when I reflect carefully upon its points one by one, it appears to be far otherwise ..."

So wrote Georgius Agricola more than 400 years ago in *De Re Metallica*, the first major attempt ever made to assemble systematically in print the world's knowledge on mining, metallurgy and industrial chemistry. The author, a municipal physician in Saxony, spent 20 years in its preparation, writing in Latin so that his work might obtain the widest circulation then possible at home and abroad. He was already dead in 1556 when it emerged from the presses in Basel, across the border in Switzerland, scarcely 100 years after Gutenberg had produced the first printed book in Europe.

For two centuries, Agricola's work was to continue to be the leading textbook for the world's miners and metallurgists. In many mining regions, including Spanish South America, it was chained to the church altar and translated by the priest to the miners between religious services. Today it remains one of the most treasured scientific classics of all times, a monument to a man dedicated to the spread of truth about one of civilization's oldest industries.

Agricola was born at Glauchau on March 24, 1494 when the world was on the threshold of the Renaissance; the Humanists had hardly begun the stimulating criticism that gave fire to the Reformation, and

Luther, who was to turn Agricola's homeland into the cradle of the great movement, was only one year old. Erasmus, later Agricola's friend and patron, had but recently become an Augustinian canon, but was already feeling the constraints of monastic life which bred in him his passion for freedom. Italy was a busy workshop of antiquarian research, of painting and sculpture, translation and study, and its universities were attracting students from the rest of Europe who were to return to their own cities, inspired by the surge of learning. At Agricola's birth, Columbus had just returned from his great discovery and it was only three years later that Vasco da Gama rounded the Cape of Good Hope.

Into the rosy dawn of this bright new world, Agricola came as Georg Bauer ("peasant"), his name being Latinized by his teachers, as was the custom of the time. When he was 20 he entered the University of Leipzig, where he gained the degree of *Baccalaureus Artium*. He taught Latin and Greek for a year or two at the municipal school at Zwickau, becoming Principal of a small staff, which included Johannes Förster, better known as Luther's collaborator in the translation of the Bible. In 1522, he returned to Leipzig University as a lecturer for two years before joining the students flooding into Italy, to study medicine and the natural sciences at the universities of Bologna, Venice and Padua, finally taking his degree in medicine at one of them. It was about this time, too, that he made the acquaintance of Erasmus, who had settled in Basel, where he published a series of his works, and—significantly for Agricola—had become editor for Hieronymus Froben and Nicolas Episcopium, one of the leading European publishers of the century and whose imprint was to be carried on *De Re Metallica*.

In 1526, Agricola returned to Zwickau and in the following year was appointed physician to the little Bohemian town of Joachimsthal, on the eastern slope of the Erzgebrige, in the midst of the then most prolific metal-producing district of Central Europe. In this bustling mining town of several thousand inhabitants there was plenty to keep him busy as a doctor—and still more to occupy his interest in the life and work of his patients. Freiberg, which had already been a mining town for 300 years, was but 50 miles distant, and the same radius included most of the mining towns frequently mentioned in *De Re Metallica*: Schneeberg, Geyer, Annaberg and Altenberg, and not very much further away were Marienberg, Gottesgab and Platten.

According to Agricola's own evidence, he spent all the time not required for his medical duties in visiting the mines and smelters; in reading, in Greek and Latin, all references that he could find by previous authors; and talking with the most learned of the local mining folk. Among these was Lorenz Berman, who became his "learned miner" in *Bermannus*, his first small book, written in dialogue form, a sort of catechism on mineralogy, mining terms and mining lore. The script was first submitted to Erasmus, who not only arranged publication with Froben, but also wrote a foreword, praising the book—and Agricola's Latin. In 1533, he published another book, a discussion on Roman and Greek weights and measures, and at about this time he started work on *De Re Metallica*.

From now on words came thick and fast. It would seem that, from the start, he had a pretty good idea of the magnitude of the task which he had set himself, and he resigned his job at Joachimsthal to devote the next two or three years to travel and study among the mines. But he had to earn a living, and he eventually accepted an appointment as city physician at Chemnitz, where he lived until his death in 1555. Beginning in 1546, he published, as a prelude to his main work, a series of books on mining, metallurgy, geology and animals in mining, as well as a small work on the Plague; he also wrote on medical, religious, political and historical subjects, reflecting his many interests and a dedicated professional outlook.

As a young man, Agricola had shown some tendency towards liberalism in religious matters and was the author of some anti-Popish epigrams, but after his return to Leipzig he never wavered and, continuing his friendship with Erasmus, remained loyal to the Catholic faith, despite the fact that he was surrounded by men who strongly supported the Reformation. When he reached 52, his broad interests became even broader; he was elected Burgher of Chemnitz, and in the same year was appointed Burgomaster by Duke Maurice of Saxony, and held the office for four terms. It is a measure of his popularity and liberalism that, as a staunch Catholic, he served a Protestant duke in a realm predominantly and militantly Protestant. He was obviously highly regarded by the Duke and, after the latter's death, by his successor and brother, Augustus. He served both dukes on many diplomatic and military missions during the wars in which they were engaged, and it was to them that he dedicated his *magnum opus*.

*

Apart from the results of his own observations, Agricola had little material on which to base his work. In this period there were really only two guides for a man in search of truth: one was the authority of his religious and lay teachers, including Biblical texts and the sayings of the ancient philosophers; the other was what an inquiring mind could note and absorb. Apart from the Book of Genesis, there had been few attempts at fundamental explanation of natural phenomena, other than those made by the Greek philosophers and the alchemists. Agricola scarcely mentions orthodox beliefs, and with the alchemists he had no patience at all. But his deep classical learning obviously coloured his thinking; to some extent he had to be a follower of Aristotle, Theophrastus and Strabo, the whole thought of the learned world at this time still flowing from these and other leaders of the Peripatetic school. Had he not, however, departed radically from the teachings of this school, his work would have been but small contribution to the advancement of science. Certain of the philosophers' teachings he refuted with great vigour, pinning his faith to the results of observation rather than to inductive speculation. In his own words: "Those things which we see with our eyes and understand by means of our senses are more clearly to be demonstrated than if learned by means of reasoning."

In describing the famous quarries of Carrara and Pentelicus that supplied marble for the great buildings of Athens and Rome, Strabo had repeated the generally-accepted idea that the voids left by mining were gradually replaced by new marble growth. Even Agricola believed in the ability of minerals to procreate and grow—he refers to the iron deposits on the island of Elba "replacing themselves" after 10 years—and, indeed, the notion was accepted by many, right through to the 19th century, and still forms part of the folklore in some mining districts of Europe.

Aristotle taught that the heat and rays of the sun, penetrating the earth's crust, made possible new combinations of elements, created metals and minerals, and formed various kinds of "stones". This belief continued into the Dark and Middle Ages and was described in some detail by Albertus Magnus in *De Mineralibus*, written about 1260 and first published in 1478. Although basically Aristotelian, there is evidence that he applied a considerable degree of personal observation, travelling widely to visit mines and study metals. Albertus was one of the most important German scientists of the Middle Ages and his book, whilst by no means free of alchemistic and mystical views, was undoubtedly the best-balanced and most comprehensive which had yet been written on the subject.

A hundred years earlier, Marbodus, Bishop of Rennes, had written a book on stones which were alleged to have some remarkable characteristics: some were claimed to cure blindness, others to prevent drunkenness; one would freeze boiling water, and another would provide protection against the pestilence. Inevitably, this proved a best-seller: some 100 manuscript copies have survived; it was translated into French, Provençal, Danish, Hebrew and Spanish, and more than a dozen editions followed the invention of printing.

More seriously, the almost universal belief in a constant process within the earth's crust which turned the baser metals into gold, was described in the *Bergbüchlein*. Published in 1497, this was probably the earliest book printed in Europe which dealt specifically with practical mining. Its author did not put his name to his work, but he is believed to be Ulrich Rühlein von Kalb who, like Agricola, was a physician (in Freiberg) and later became Burgomaster. The book describes seven metals and their ores, methods of prospecting and the tools needed in mining. The text is attractively presented in the form of questions, put by the apprentice Knappius and the answers given by the mining expert Daniel. *Bergbüchlein* represents the first serious attempt to set down some of the fundamentals of mining which, until then, had been transmitted by word of mouth from generation to generation, for the benefit both of those engaged in the industry and the general public. It bridged the gap until Agricola; by 1535, by which time work on *De Re Metallica* had begun, seven editions of *Bergbüchlein*, all in German, had appeared and the book continued to be in demand throughout the following century. Only two copies of the original edition survive: one in the Municipal Library in Augsburg and the other in the Bibliothèque National in Paris.

Another anonymous work was also in circulation at this time. Entitled *Probierbüchlein*, it was essentially a collection of papers for the mining assayer, and this, too, was frequently reprinted. A few years earlier the first printed German mining regulations had appeared, based on rules laid down for operations at Schreckenberg.

In Italy, a significant contribution came from the metallurgist and armament maker, Vannoccio Biringuccio, who described the process of extracting metals and the preparation of chemical substances in his *Pirotechnia*. Biringuccio enjoyed the patronage, and suffered the vicissitudes, of the Petrucci family, one of the most powerful in Sienese politics, and became head of the foundry of Pope Paul III. Unlike so many of the obscure alchemical writings which circulated in his country, his work gave clear, practical instruction; it quickly became a standard reference book and today remains a valuable source of information for historians on the state of chemical knowledge of the time.

Finally, among the published works which preceded *De Re Metallica*, must be mentioned the *Cosmographia* of Sebastian Münster, which included a considerable section on mining and, from the 1550 edition onwards, illustrations of operations below and above the surface. In his treatment of the subject, Münster used the early works of Agricola as one of his principal sources of information—and paid due acknowledgement to the fact—but his huge volume, which attracted the educated world of the day and was warmly received by both Catholic and Protestant because of its impartial tone, was to overshadow *De Re Metallica* to some extent, if only because of its international distribution: in the course of the next 100 years it went to 49 editions in six languages—German, Latin, Italian, French, English and Czech. In fact, the *Cosmographia*, with its portraits and drawings, is an early example of how a popular work of science can spread far more knowledge than a textbook because of the wider circle of its readers.

Meanwhile, patiently, year after year, Agricola stuck to his task, travelling, observing, discussing and meticulously setting down details of every mining and metal-processing activity and practice that he considered of any value. Slowly, he unravelled the mysteries of the "secret art", many processes of which had until now been held by families and guilds, or other closely-knit societies. He completed his manuscript in 1550, but did not send it to the press for another three years, and two further years were to elapse before publication. Chief reason for the delay was the laborious preparation by a team of artists of the illustrations, which the author considered of great importance, "lest descriptions which are conveyed by words should either be not understood by men of our times, or should cause difficulty to posterity". His insistence on this final perfection meant that he was never to see his book in print, but the many intricate and expressive woodcuts which adorn the pages were to add greatly to its attraction and are today perhaps more widely known than the text itself. This first Latin edition is a tall folio of 586 pages, including indexes. The colophon states that it was finished in March 1556 at Froben's press in Basel. The same woodcuts were used in subsequent issues. In the following year Froben arranged a translation into German, and Philipp Bechius, professor of philosophy and medicine at the University of Basel, was chosen as translator. Unfortunately, he made a poor job of it and the text was not corrected until 1928 when the Agricola Society of the Deutsches Museum issued a new translation from Berlin, prepared by a committee of eight scholars. An Italian translation was made by Michaelangelo Florio, a Protestant refugee from Florence and teacher of Italian in London; it was published by Froben in 1563 and dedicated to Queen Elizabeth. In all, four Latin editions were printed between 1556 and 1657; five German editions between 1557 and 1953; the Italian edition in 1563; and the English in 1912, reprinted in 1950.

*

It seems incredible that the world was not able to read *De Re Metallica* in English until more than 350 years after its first publication. A few attempts at translation were made, but none successful, and it was not until the early years of the present century that the task was completed by Herbert C. Hoover, son of a village blacksmith in Iowa, who, after a distinguished and adventurous career in mining, was to become the 31st President of the United States. His wife, Lou Henry, shared the work—and shares the credit for a remarkable piece of literary skill and ingenuity.

The Hoovers had first met at Stamford University where they were fellow students. Herbert, who lost both parents before he was nine and was brought up by an uncle, was sent to Stamford to study mining engineering under Dr. John C. Branner, to whom the English translation of *De Re Metallica* was dedicated. Lou Henry was also born in Iowa and, because of this, Hoover felt some obligation to assist her in her geological studies, though he later admitted that this self-imposed responsibility was stimulated by her "whimsical mind, blue eyes and a broad, grinnish smile that came from an Irish ancestor". They married in 1899, after Hoover had gained some valuable experience in Colorado, and later in Australia, where he worked for Bewick, Moreing and Company, of London, which had already been managing mines all over the world for more than 150 years. After their wedding in California, the couple went directly to Peking, where Hoover became chief engineer of the Chinese Engineering and Mining Company. Here, they were caught up in the Boxer Rebellion and lived through the siege of Tientsin by anti-foreign Chinese. Later, in 1901, Bewick, Moreing offered Hoover a partnership and for the next seven years he travelled extensively on their behalf, circling the globe five times. By the time he was 34 he had acquired wealth and a world-wide reputation in his profession as well as interests in a number of mining companies. In 1908, he started his own consultancy which he set up in London, with offices in New York, San Francisco, and subsequently in Petrograd and Paris.

Always a scholar, Hoover had become much interested in the older literature of engineering and applied science generally, and had made a considerable collection of 15th and 16th century books. *De Re Metallica* fascinated him; its translation presented a challenge which he could not resist. But he could not tackle it on his own. As he admitted in his memoirs, "My own study of Latin had never gone beyond some elementary early schooling and a few intermittent attempts to penetrate further into that language and literature after I left college. Mrs. Hoover was a good Latinist . . . after she brushed up a little we found we could work it out and finally, in 1907, we resolved to translate it jointly."

It was not, of course, as simple as it sounds. There were formidable difficulties for, whilst Agricola's Latin was scholarly enough, he was writing of mining practices and terms for which there were no Latin words or phrases: the language had been dying for a thousand years and had failed to embrace the technology and commerce of the times. It would have made the Hoovers' job easier if Agricola had used colloquial German terms to describe these technical operations and equipment, but instead he coined or adapted several hundred Latin expressions to cover them. Every one of these had to be carefully worked out. Even then the problem was not solved, for many of the processes described by Agricola had no counterpart in mining practice in the English-speaking world and therefore had no equivalent expression in English. Moreover, the original German terms themselves had suffered change and obsolescence with the passage of time, and some, indeed, had changed their meaning in the 20 years that Agricola spent in preparing the text.

The task therefore required as much scientific detective work as translation. The Hoovers grappled with the manuscript, sentence by sentence, during their spare time for over five years. They lugged it all over the world for the odd moments that might be available to work on it in foreign capitals, in lonely mining camps, on ships and in trains. Much of the translation was done in the Hoovers' London home, a delightful old house in Campden Hill, so old in fact that despite its situation not much more than a mile from Hyde Park Corner, its original tenants were required to "prevent their cows from wandering in the High Street" and to "refrain from hanging laundry in view of neighbours". Here, the Hoovers worked in the oak-panelled library, with its fine fireplace and leaded-glass bookcases, the table piled high with manuscripts and reference books, Hoover at one end, his wife at the other, the children clambering onto their mother's lap, demanding to know what the book was all about.

By 1912, the translation was completed to their satisfaction, together with thousands of words of explanatory notes. There had never been any thought in the translators' minds that their work might be of commercial value: it had always been a "labour of love" and, on completion, their sole desire was to present it to the mining engineering profession. To this end, they consulted their friend Edgar Rickard, who was then business manager of *Mining Magazine*, which had been founded in London a few years earlier, largely on the initiative of Hoover and some of his associates, and edited by Edgar's cousin, T.A. Rickard. Edgar knew of a commercial printer in the Midlands, named Alfred Frost, who had a love of old books. When Frost saw the manuscript and the original, his eyes glistened; all his life he had wanted to do such a book. He offered to print it at bare out-of-pocket costs, and *Mining Magazine* contracted to publish it. To make the book look as much like the original as possible, Frost found a paper-maker who could produce a 16th century linen paper. He had a fount of type cast in exact reproduction of the original, except for the ancient letter S, confusing to modern readers, and printed the book as a folio volume with all the old prints and illustrations. "It was", recorded Hoover, "in astonishing fidelity to the original. By a little juggling of the opening sentence of each chapter, we were able to include repro-ductions of the old illuminated initial letters. Mr. Frost bound it in white vellum as the original was bound. No more beautiful job of printing was ever turned out."

Three thousand copies were printed and more than half of these were sent as gifts to engineers and institutions. To ensure that other members of the profession could obtain copies, 1,000 were placed on sale through *Mining Magazine* at a nominal price which hardly covered the binding.*

The English version of *De Re Metallica* was not just a book; for the mining world, it was an event. Four centuries earlier, publication in the original Latin represented the first major contribution which

---

* The price was 21 shillings. Few copies change hands today at under £200.

literature had made to the development and understanding of the industry. Now, at last, the wisdom and philosophy of Agricola, acknowledged "father of mineralogy", was made available for all, serving, in the words of the translators, "to strengthen the traditions of one of the most important and least recognized of the world's professions".

CHAPTER TWO

The Years Between

Herbert C. Hoover

The mining methods and processes described by Agricola in *De Re Metallica* had long since been superseded when the Hoovers made their English translation in the early part of the present century. The translators never claimed that their work was of any "practical" value; they believed only that they had cleared away the undergrowth which had obscured an important milestone in the development of one of the most basic of human industrial activities, a milestone which they felt was "far more worthy of preservation than the thousands of volumes devoted to records of human destruction".

The Hoovers made Agricola famous and, in doing so, added stature to the industry and to the men and women working in it. To Herbert Hoover this obviously meant a great deal. A few years before he decided to tackle the translation of Agricola, he had given a series of lectures at his old university at Stamford, and at Columbia University. These lectures were subsequently brought together in a small textbook under the title, *Principles of Mining*, which was reprinted many times, and for years continued in use in the engineering schools. Apart from its technical instruction, the book underlined the role of the engineer as an economic and social force, the creator of new industry, new jobs and new standards of living. Hoover was always anxious to put the profession's functions and responsibilities into perspective, remembering perhaps, with wry humour, an incident that had occurred during one of his many voyages across the Atlantic. On this occasion he had shared a table with an English lady "of great cultivation and happy mind", who contributed much to the conversation during the days at sea. At the final breakfast, as the ship entered New York harbour, she turned to Hoover and said, "I hope you will forgive me, but I should like to know, what is your profession?" Hoover told her, whereupon she appeared quite startled. "Why", she said, "I thought you were a gentleman." Little wonder that Hoover took every opportunity to endorse Agricola's own assertion that mining was "a calling of great dignity".

Whilst *De Re Metallica* has been of little more than academic interest to the 20th century, it had great impact on society when it first appeared 400 years earlier. Hoover himself observed: "There is no measure by which we can gauge the value of such a work to the men who followed in this profession during centuries, nor the benefits enjoyed by humanity through them. Science is the base upon which is reared the civilization of today and, whilst we give daily credit to all those who toil in the superstructure, let none forget those men who laid its first foundation stones." That the book passed through 10 editions in three languages, in a period when the printing of such a volume was a major undertaking, is in itself sufficient evidence of the importance in which it was held; it is a record that no other book on its subject has since equalled.

Literature contributed little to the knowledge of mining before *De Re Metallica*—and little of much consequence for 200 years after its first publication. Conrad Gessner, a contemporary, and admirer of Agricola—a Swiss and another physician—was a prolific writer on plants and animals; he later became interested in fossils and stones and was the first to classify minerals by form rather than by name. The direct influence of Agricola was also reflected in the writings of Lazarus Ercker, chief superintendent of mines in the Holy Roman Empire and the Kingdom of Bohemia, who in 1574 published a treatise on ores which was basically a textbook for the assayer. Of far greater importance was the influence which Agricola's work had in encouraging, at Freiberg, a central source of mining and metallurgical information, which was later to be formalized into a definite curriculum and in the establishment in 1765 of the famous Freiberg Mining Academy.

The 17th century is sometimes described as an age of planning but, if there were significant advances in mining technology, they were not reflected in its literature. The works of Agricola, Biringuccio and Ercker remained pre-eminent and most of the new publications were little more than a plagiarism of the past. A few names survive in the records: for example, Georg Engelhardt, a mining foreman at Brunswick, is credited with an anthology of some repute; and in 1674 Christian Hoffmann was the author of a little book on gold mining in the Reichenstein mountains, perhaps the most interesting feature of which was a comprehensive index of mining terms, which ran longer than the descriptive text.

However, if for nothing else, the century was notable for the appearance of some early English books on mining. Sir John Pettus, Deputy Governor of the Royal Mines (he was also an M.P. and Deputy Lieutenant for the County of Suffolk) produced in 1670 a record of "the history, laws and places of the chief mines in England, Wales and the English Pale in Ireland". Pettus also translated into English Lazarus Ercker's treatise, and in 1688—the year William of Orange landed at Torbay to assume the English throne—came Thomas Houghton's *The Compleat Miner*, with accounts of the "laws of the miners in the Forrest of Dean" and "the ancient laws of the miners in the King's Forrest of Mendipp".

And from France, the first feminine touch in mining literature: in *La Restitution de Pluton* (1640), Baroness Martine de Bertereau urged the exploitation of the country's undeveloped mines as a means of creating industry and employment—and of increasing the royal sources of income. Eight years earlier, the Baroness—a one-woman pressure movement—unsuccessfully petitioned the King and the State Council on the same subject, and her book was obviously designed to arouse public interest in her campaign.

With the 18th century came the age of reason. The decline in belief of the mystical, and the steady growth of a rational, intellectual approach to human and scientific problems, led to much new knowledge in the field of chemistry and physics, two of the basic disciplines in the orderly development of metallurgy and mining. Many outstanding technicians and inventors filled the stage to prepare for the Industrial

Revolution which, in the following century, caused fundamental changes in the social, political and economic structure of the civilized world. The most significant invention of the century was the steam engine and, initially at any rate, its most important application was in mining. James Watt arrived on the scene when the mining and metallurgical industries were expanding rapidly and when there was urgent need for an efficient machine to pump water from the mines. The steam engine progressively replaced human and hydraulic power and paved the way for the huge increase in the output of ores and coal demanded by the incipient industrial age.

The systematic thinking and research which characterized the 18th century found reflection in its literature; not only did many more books and scientific papers become available, but attempts were now being made to arrange the rapidly increasing amount of scientific material into some sort of order. One of the leaders in this endeavour was Abraham Werner, who entered the mining school at Freiberg as a student and later, as teacher and inspector, devoted 40 years to its development as one of the great centres of scientific learning in Europe. What Agricola had done for Freiberg in the mid-1500s, Werner continued in the second part of the 18th, and through into the 19th, century. (It was he, incidentally, who first gave Agricola his title of "father of mineralogy".) Werner imbued the school with a spirit of investigation and critical analysis previously unknown in the geological field, and his collection of books constituted the best mining library in the world, his own work on minerals, published in 1774, being not the least important of the many volumes on the shelves.

It was during this period that a start was being made to classify minerals by chemical composition rather than by their physical qualities. Chemistry, in its modern sense, was developing as a result of the investigations of the many-sided genius, Lavoisier, but it was not until the beginning of the 1800s that the chemical analysis of minerals was sufficiently advanced to be useful to mineralogists. Werner was an ardent advocate of the theory that water was the sole agency in the deposition of the successive layers which constituted the earth's crust, heading the school of "Neptunists", as opposed to the "Plutonists", led by James Hutton, who held that fire, as well as water, played a significant part in the process. Werner erred in his teaching that granite was a sedimentary rock; Hutton argued otherwise and although his *Theory of the Earth*, which appeared in 1785, was first received with scepticism, it gradually became the accepted explanation of the igneous nature of the earth's crust.

It was controversies such as this that stimulated the development of a new form of presentation of scientific thought and opinion. The middle of the 17th century had seen the beginning in several European countries of a number of periodicals, learned in character and international in scope, such as the French *Journal des Savants* (1655), *Philosophical Transactions* (1665), *Geornale de Letterati* (1668), *Miscellanes Curiosa* (1670) and several others. These encyclopaedic journals published a few pages of reports of discoveries and inventions, general scientific observations, and book reviews.

With the age of reason some 50 years later, came a spate of weeklies, whose main object was to raise the social and moral level of society, though a few devoted a little space to technical matters. Of greater importance to the scientist and technician were two kinds of journals which became popular in the second half of the 18th century. The first of these were the so-called intelligence sheets, which were also a medium for the promotion of the arts; the other group consisted of regularly-produced magazines which placed special emphasis on the sciences and dealt at some length with technological questions. They included the *Hamburg Magazine* (1747), *La Feuille Nécessaire* (1759), *Journal Anglais* (1775) and the *Leipzig Magazine* (1781).

For the most part, these periodicals did not contain original reports, but interpretations from different sources, providing the sort of background reading to economic, political and cultural events that the quality weeklies publish today. Towards the end of the century more specialized journals began to appear, at first in the scientific, and later the technological, field. The first to be devoted to mining and smelting appears to have been the German *Magazin Bergbaukunde*, published in 1785 by Johann Friedich Lempe, followed two years later by *Bergmännische Journal* from Köhler and Hoffmann, which was succeeded by *Neue Bergmännische Journal* from the same publishers in 1795.

None of these papers survives today. The oldest mining journal in the world which is still in production is the French *Annales des Mines*, which began life in 1794 under the title, *Journal des Mines*, whilst in the English-speaking world, the London *Mining Journal* has been in continuous publication since 1835.

The apparent paucity of mining literature in Britain before the 19th century is a little puzzling in view of the country's long mining traditions. The links stretch back through history to the Bronze Age when the first explorers, who came from Mediterranean lands, discovered that within the mixture of sand and stone, washed down by the streams from the granite hills of Cornwall, was a precious ore, tin which, when mixed with copper, enabled them to forge sturdy weapons and tools.

One explanation which has been offered is that whilst, in continental Europe generally, the state promoted mining and was anxious to keep its progress under constant review, most mining and other industrial activity in Britain was the result of private enterprise. Much of this enterprise, it has been argued, was promoted by people with strong Calvinistic backgrounds, men who were entirely absorbed in the practical aspects of their endeavours and had neither the obligation, nor the inclination, to account for their work in books. Even in the 1800s, there were many more English books of popular science, directed to the public at large, than specialized works for the technician.

Similarly, the 19th century literature of the United States contained few references to mining. Early American mining relied on publications from Europe and, even at the time of Robert Peel, who was Professor of Mines at the Henry Crumb School of Mines at Columbia University and compiler of the *Mining and Engineering Handbook*, the textbooks of foreign authors were mostly used. Independently-produced American mining literature was not very much in evidence until the beginning of the present century, following closely the establishment of mining schools and institutions in the latter part of the 1800s.

*

It is more than a coincidence that *Annales des Mines*, which has survived more years than any other mining journal, was born during

17

the French Revolution, at a time when France was not only being torn asunder in the struggle for "liberty, equality and fraternity", but also when the country was heavily engaged in a series of wars with one or more of the other European powers in an attempt to impose on its neighbours the fundamental changes which it had made in its own state and society.

The paper, originally called *Journal des Mines de la République*, was formed on the initiative of the Committee of Public Safety (*Comité de Salut Public*), which had been set up after the fall of the Girondins, and only a few months after Louis XVI had been taken to the Place de la Revolution—the modern Place de la Concorde—and there executed amidst cries of "vive la nation!"

The formidable Committee, dominated alternately by Danton and the "incorruptible" Robespierre, soon succeeded by hard work and efficient centralization, in exercising control over the ministries and administrative bodies, the judges, generals and *commissaires* of the public, whom it appointed, purged or dismissed at will. Among the ministries critically examined was that concerned with the administration of the country's mines and mining schools. Drastic changes were made in the interests of the new regime and, in order to ensure that the people were kept fully informed and the mining profession adequately supplied with the best technological information and opinion, the Committee decided to publish a monthly magazine. Events moved quickly in France in those days and before the first issue, in October 1794, both Danton and Robespierre had themselves followed the many thousands of victims of the Terror, which they helped to create, to the guillotine.

It was hardly the atmosphere in which an enterprise in class-journalism could be expected to flourish. The authority of the Committee was absolute, resting ultimately upon its ruthless use of force; its travelling agents were never slow to exercise their power to arrest and send for trial all who obstructed the government, or who were considered to oppose its policies in any way.

But the magazine survived. Its first issue proclaimed that France was as rich underground as it was fertile on the surface. It deplored that the country had in peacetime been paying millions of francs to foreigners for raw materials which could be mined within its own borders; it announced that a new department of minerals was to be set up, with libraries, laboratories, a bureau for planning and design, and a new system of technical instruction and education.

Four years later—by which time the Directory had replaced the Committee of Public Safety as the effective government of the country—the Journal reported speeches at the opening session of l'Ecole des Mines, showing that France had already become self-sufficient in steel and that good progress was being made in the development of other industrial activities. There were, for example, 2,000 furnaces for the production of iron and steel; 400 coal mines were in operation and there were 200 more potential producers; silver, lead, copper, manganese and cobalt salts were being mined; and the springs of the eastern provinces were supplying large quantities of common salt. Moreover, the conquests of the French armies had acquired for the Republic rich mines of mercury, copper, zinc and coal. The only metal still lacking was tin, but there were signs that resources might be discovered in the heart of the country.

During the ceremonies which marked the inauguration of the school, tribute was paid to the Journal's editor, Citizen Coquebert, of whom it was said " . . . the Journal profits greatly from the studies which this learned man has made in foreign languages, and from his talent to write about science . . . the Journal is the depository of knowledge concerning all the new discoveries . . . it reports on the state of our mines and factories and gives the lie to all those who would challenge the abundance of our resources."

The early issues of the Journal, under Citizen Coquebert's direction, did in fact lay down an editorial policy and established patterns which have changed but little during the paper's history. The Journal's *raison d'être* was stated in the first number to be to reproduce "in entirety or in extract" the most important scientific works published both at home and abroad. Throughout its life the magazine has taken a much broader view than most similar publications in other countries; it has never confined itself wholly to mining and metallurgy but has, when appropriate, published studies on the development of chemistry, physics, railways, steam engines and the motor car, and, more recently, problems relating to fuels and power, nuclear energy and the environment. And although today the Journal is specially concerned in publishing studies made by the engineers of the Administration des Mines and by professors and lecturers of its mining schools, it still calls upon technical writers in many other countries to continue to give it the international character and appeal which it established more than 180 years ago.

In Paris, the yellowing file-copies of *Annales des Mines* are bright with the names and achievements of those who brought about a great awakening of science in one of the most remarkable periods of the world's history. Its articles deal with practically every scientific and technological development of significance by men such as Leblanc, Lavoisier, Priestley and Cavendish, who laid the foundations of modern chemistry; André-Marie Ampère, François Arago and Hans Oersted in the field of electro-magnetism; Humphry Davy, inventor of the miner's safety lamp; de Saussaure, the Swiss physicist and early Alpine explorer; the British scientists, William Hyde Wollaston and John Dalton; Matthew Boulton, the Birmingham manufacturer who introduced James Watt's steam engine; Robert Fulton, the American engineer who made the first submarine and pioneered the development of the steamboat; and René Hauy, the French mineralogist who, by accident, became the founder of the science of crystallography. Contributors also included Claude Louis Berthollet, who instructed Napoleon in chemistry and accompanied him on his expedition to Egypt in 1798; his achievements were many, though his attempts to produce a substitute for saltpetre, which then had to be imported for the manufacture of gunpowder, resulted only in a brilliant display of coloured fireworks!

One other name, among many, catches the eye in the Journal's files, that of Dieudonné Dolomieu, the French geologist and mineralogist, who gave his name to the mineral, dolomite. Admitted in his infancy to the Order of the Knights of Malta, he killed a brother knight in a duel, was condemned to death but, in consideration of his youth, was pardoned after nine months in prison, which he spent studying the natural sciences. He made an exhaustive study of the Alps, and in 1791 described the mineral dolomite, to which he gave his name.

Captured on his way home from Napoleon's Egyptian campaign, he was imprisoned in Messina, in a pestilential dungeon. Forbidden writing material, he made a pen from a piece of wood and, with the smoke of a lamp for ink, wrote his treatise, *Sur la philosophie minéralogique et sur l'espèce minérale* (1801) on the margins of a Bible, the only book he possessed.

Apart from a gap of two years between 1799 and 1801, and during the First World War, *Annales des Mines* has been in continuous production and has remained an official organ of the government's mining department despite a multitude of poltical changes, surviving Monarchy, the First and Second Empires and the various Republics. Its format has changed several times, and in various periods it has been published monthly—as it is today—every other month, and quarterly. It changed its name to its present title in 1860, but it has never changed its character, nor has it departed from the high standards which Citizen Coquebert set for it when the first issue appeared in those days of bloody revolution when the wagons rumbled daily through the streets of Paris to the guillotine. It is unique in mining journalism—and indeed in the whole history of technical publishing.

CHAPTER THREE

One Man and his Pen

Lord Ashley pleads the miner's case

The circumstances in which mining journalism was born in Britain in the early part of the 19th century were very different from those in which *Annales des Mines* had been established in France some 40 years earlier. Now, a series of gigantic wars had at last come to an end. Napoleon had failed in his attempt to destroy British trade. After Elba, he had lost the magic of victory; at Waterloo in 1815 the French threat was finally removed and a period of national peace and prosperity lay ahead. Recovery in Britain was slow, and sometimes painful. Expenditure out of all previous conception had been poured into the war effort, but industry soon began to send up new shoots, encouraged by inventions and discoveries in almost every field of endeavour. It was the age of Cavendish and Faraday, of John Dalton and Humphry Davy, of George Stephenson and the *Rocket*, and a host of other scientists and engineers whose work was to affect the lives of everyone in the land.

Equally important, it was also an age of much-needed social reform. At the beginning of the century the working people lived under heavy burdens; fettered by vexatious laws, they were little more than a part of the machinery which they operated. To organize in their own interests was a crime against the Combination Laws; the factory system was cruel and oppressive and employed mostly women and children. The criminal laws were brutal almost beyond belief; there were few, if any, countries where so many actions were punishable by death; more lenient sentences were transportation—mostly to Australia—public whipping, the stocks and the pillory, which was not abolished until 1837, the year Queen Victoria came to the throne. In theory, there was Press freedom, but publishers remembered that less than 50 years earlier reporting debates in Parliament was a breach of privilege which carried dire penalties; and as recently as 1811 the editor of a country newspaper had been sent to prison for criticizing the continuing practice of military flogging, citing the case of a soldier who had been sentenced to receive 1,000 lashes, 750 of which were actually administered.

British mining was on the threshold of its most spectacular development. Of copper, a chronicler of the period wrote: "The quality of ores produced from the Cornish and Anglesey mines, the latter in particular, was so large, the profit so ample, the rise of modest men into nobles so swift, and the increase of wealth by those who were the fortunate holders of a comparatively few acres of copper ore, so vast that it produced the natural result—a rush into the copper trade." By the middle of the century, Devon and Cornwall constituted the largest copper producing district in the world. The mines, which had originally been concentrated in the west, now sprawled eastwards through the county, and the great Caradon mine, set on two hills north of Liskeard, alone employed more than 4,000 people.

The collieries were hard pressed to meet the demands of the Industrial Revolution, which showered its sparks over the new communities that had sprung up around the factories and workshops. In a country where timber for fuel was scarce, coal was indispensable for both domestic and industrial purposes. The iron trade depended on its plentiful supply; without coal, the new steam engines would stand idle; coal mining was a major factor in the national economy and a big source of employment—though the manner in which many of the mines were worked remained a scandal which took many years to eradicate.

The birth of the *Mining Journal* in 1835 was therefore no accident: it was a product of a time when British mining, and industry generally, demanded not merely a chronicle, but a critical analysis of its conduct and development. In many ways it was long overdue. Explaining its *raison d'être*, an editorial in the first issue remarked: "If we were alone to consider the importance of our metallic products, iron, copper, tin, lead, and other metals, it becomes a matter of astonishment that no journal has ever been published to record discoveries and the results of labour which might, if registered, have been productive of so much advantage to the interests of society; while it is equally a matter of surprise that, in the absence of any practical work of reference from which information might be collected, we should have arrived at that high station in scientific acquirements to which we may so justly lay claim."

The foundations of the paper were laid a few years before that editorial was written, by a young man, barely in his twenties, who had already published a "Compendium of information relating to companies formed for working British mines", and a comprehensive glossary of mining terms. His name was Henry English, a name which is inextricably bound up with the early history of technical and trade journalism.

About 1830, English began publishing—somewhat irregularly, it is true—a quarterly review of mining which paved the way for the establishment of the *Mining Journal and Commercial Gazette*, the first industrial paper to be produced in Britain. Its foundation was remarkable if only for the fact that it appears to have been entirely a one-man enterprise. No evidence has been found in its records of any subscriptions from outside sources for its capitalization and it was with obvious satisfaction that, some 10 years after its inauguration, English himself affirmed that he had been supported "neither by communications nor subscriptions", adding: "It is only to perseverance and making friends, by a strict adherence to principle, exposing abuses, and upholding undertakings honestly taken up and which held forth promise, that we have maintained our position." This unity of direction and control undoubtedly contributed to the individuality and independence which the Journal has consistently displayed, and today it remains a private undertaking, serving the industry generally without allegiance to, or bias towards, any one section of its interests.

The first issue, dated August 29, 1835, was a broadsheet, a popular format of the day, 11 in. wide by 17 in. deep, and this was to remain its shape for more than 70 years, the number of its pages varying to accommodate the volume of available news and advertising—and to accord with economic expendiency.

Its "prospectus", as set forth in its first editorial, was essentially realistic and underlined the strong economic sense of the founder. English emphasized that mining was not only a matter of interest to the man of science, but to the capitalist and trader. Engineering, geology, mineralogy, chemistry, physics—these were all subjects which it was natural for a "mining" journal to cover; but the second part of the title, *Commercial Gazette*, reflected the editor's determination not to be inhibited by too closely-defined parameters. *Annales des Mines* had already adopted an "open" editorial policy and, whilst the Journal never covered quite such a wide field as its French contemporary, it fully recognized the potentialities of a readership which would include, for example, those interested in the obvious relationship between mining and the railways, which were then on the brink of enormous development. Indeed, at a later date, the paper was restyled *Mining Journal and Railway and Commercial Gazette*, the second part of the title being dropped only after the Second World War.

English had a special interest in the railways as a means of improving communications between the mining centres and speeding up the flow of news to his office in London. He had already started his *Mining Review* when the Manchester and Liverpool Railway, the first to be used for passenger traffic, was opened in 1830 with a grand procession of eight locomotives—all built by Stephenson on the model of his famous *Rocket*. In the early days of his editorship, whenever English needed information he had to go and gather it himself, and this meant stage coach trips to the mining districts in winter as well as in summer. On one occasion a drunken driver upset the coach, and he had to spend several weeks on the road, recovering from three broken ribs. Perhaps, therefore, more than anyone, he welcomed the new means of transport, despite the warnings of a contemporary periodical that passengers who were to be "whirled at a rate of 18 or 20 miles an hour" ran the risk of being "scalded to death, drowned by the bursting of the boiler, or dashed to pieces by the breaking of a wheel".

In view of the considerable difficulties in obtaining "hot" news, it is not surprising that much of the editorial content of the early issues of the Journal was concerned with the reports of the big mining companies. Indeed, in one issue, English found it necessary to apologize for the fact that, in narrating the operations of the public companies, original contributions and information had to be severely limited. However, even if English could be held to have placed undue emphasis on company reports, this certainly did not preclude him from expressing sharp criticism of the mining companies whenever he thought it justified. He was quick to denounce abuses in management, and was particularly severe in exposing unsound or fraudulent enterprises—of which there were more than a few at this time—and he became involved in several libel actions with leading city magnates. Most of these actions appear to have been withdrawn by the complainants before they reached court, but from those which did go for judgment, English invariably emerged triumphant, though, win or lose, such cases must have cost the paper a great deal of money.

*

In one of his first leading articles in the Journal, English made reference to what was to become the great mission of his life—a reduction in the appalling rate of accidents in the British mines, which he sought to achieve by the introduction of safety regulations, enforceable by law, coupled with the establishment of a proper relief organization for the families of the victims of mine disasters. Many of the accidents of which English complained arose not only from the neglect of mine-owners to ensure reasonable safety, but from the fact that in these early Victorian days, as many women and children as men were employed in the mines—certainly in the collieries—and few of them had any form of training or instruction. Children were often put to work from the age of five or six, spending 12 to 14 hours a day underground in dreadful conditions.

Champion of the movement to get this state of affairs changed was Lord Ashley, later Lord Shaftesbury, who was also passionately devoted to campaigning for better conditions in the factories and for the abolition of the employment of "climbing boys" by the chimney sweeps. In May 1842, the Royal Commission on Children's Employment, which had been set up on Ashley's initiative two years earlier, issued its report on the mines. Among the commissioners who signed the report was Dr. Thomas Southwood Smith, who had the foresight to realize that many people who would not read an official report might well turn over the pages to glance at any pictures that it might contain. He therefore persuaded his fellow commissioners to have the report illustrated with some clear, simple line-drawings. It was an imaginative decision, a lesson in how the graphic presentation of the most wordy report can compel its attention. When these pictures first appeared the effect was electric, and a wave of horrified indignation swept the country. Among the drawings were some which showed little girls carrying huge baskets of coal up long, steep ladders; five and six-year-olds sitting alone in the darkness, opening and shutting ventilation doors; and women and children, harnessed to trucks by chains which passed between their legs, crawling on all fours to haul loads of coal through tunnels less than two feet high.

The public was even more shocked by a drawing of a half-naked boy and girl sitting face to face astride a bar, she holding a rope and he with his arms round her body, whilst a mis-shapen old woman turned a windlass to lower them down the shaft. Characteristic of the high moral sense of the gentlefolk of the time, it was not so much the obvious danger of this mode of descent, but its "obscenity", which troubled the Victorian conscience. Brutality could, in certain circumstances, be excused, but immorality never.

Several witnesses interviewed by the Commission declared that the colliery-owners were quite unaware of the state of affairs in their own pits. In her biography of the Seventh Earl Shaftesbury, Georgina Battiscombe remarks that "this ignorance was not altogether to be wondered at, remembering the only available means of descent; Lord Londonderry, for instance, could hardly be expected to be lowered in a small, open bucket, or to share a seat on a wooden bar with a half-naked colliery lass."

It was significant that Ashley's Ten Hours Bill to regularize work in the factories took 14 years to become law, but his Mines Bill passed through both Houses of Parliament in as many weeks, although difficulties inevitably arose in the House of Lords, where the mine-owners were powerful. It was a classic case of the pen of the artist being greater than the oratory of the politician.

But neither the Act nor the publicity stopped the accidents. They continued with frightening regularity. By 1844, English estimated that deaths were running at a rate of 2,500 a year—it had to be an estimate as not all fatal accidents were reported and no accurate records were kept by the authorities. Every issue of the Journal carried accounts of disasters—through fire-damp explosions, rock falls, drownings, breaking of the winding ropes (still mostly of hemp, the introduction of wire ropes being the subject of much lively controversy in the paper's correspondence columns), inefficient safety lamps and misuse of lamps. Coroners' inquests laconically returned verdicts of "accidental death", with no blame attaching to mine-owners and no legal action taken against those who were shown in court to be guilty of negligence.

The year 1844 was scarcely a few days old when 12 lives were lost at a colliery near Cardiff, "adding", as English wrote "another to the fearful items on the list now in course of preparation with a view to being presented to the House of Commons by Lord Ashley. Fully assured are we that no good will ever be achieved until the protection of the life of the collier and miner becomes a legislative measure and compulsory."

In this case, as in many others, the cause of the accident was a fire-damp explosion. The coroner's inquest disclosed "the neglect of the overman, whose business it was to go into the pit every morning at 3 o'clock to examine all the headings and stalls. He went down on Monday morning, accompanied by a little boy, to whom he imprudently entrusted the task of exploring a particular stall. This duty the little fellow did but partially perform, and in this particular stall it was found that foul air had generated, which was the cause of the melancholy and disastrous consequences."

At the inquest the usual verdict of "accidental death" was returned. The Journal reported: "On the delivering of this verdict the coroner adverted in terms of strong reprehension to the conduct of the overman, through whose neglect the calamity occurred, and left it to the jury to say whether he ought to be committed on a charge of manslaughter. The jury intimated that such a charge would receive no countenance from them, and the proceedings terminated."

English hammered home this "callous neglect of justice" in a leading article: "If this in itself is not sufficient to call for Parliamentary legislation and interference, we know not what is. We once again call on those whom humanity would prompt to lend their aid. If that a committee be formed, it shall have our best assistance; if we are left singly to fight the battle, we shall do so."

A week later, six more accidents were reported, including one at Duke Pit, Whitehaven, where 11 men were killed as a result of "a dreadful explosion of fire-damp". An "open" lamp was the cause. In .his leader, in which his passion sometimes affected his prose, English demanded: "How long is this to be continued? Surely, evidence to warrant—nay, demand—Parliamentary inquiry has been adduced, and we trust that the first week of the ensuing session will not pass by without the subject being brought to the attention of the Legislature. It will not be our fault if such be not the case, but how lamentable it is that in a work of humanity we do not find support. There are, we allow, one or two honourable exceptions, but we regret to say that the colliery-owners, and even the magistracy of the numerous colliery districts, are dead to the voice of humanity—to the claims of sympathy—even as those whose corpses, blackened by the fearful death to which they have been subjected, and whose families are left to mourn their loss, while the community is charged with their support . . ."

In the following week English sat down and "addressed upwards of 150 letters to parties, amongst whom are the highest in the realm, who are, or who should feel, interested in the subject", and on February 10 he jubilantly wrote: "It is most gratifying to be able to announce that the subject of Accidents in Mines and Collieries has at length been taken up by parties whose position and influence must assure success. It would be premature were we to give the names of the noble lords, members of Legislature, coal-owners, iron-masters, or those connected with mercantile affairs, who have expressed their willingness to lend their aid, while the very nature of the inquiry must ensure the support of every man who has the common feelings of humanity. We are proud to say that the day is not far distant when the blot will be removed, which cannot be considered otherwise than as a stain upon this country ... the present measure will, we trust, not only have the desired end of causing some legislation to be passed, but will at the same time form a connecting link between master and man, not easily to be severed, and the strength and value of which will, we feel assured, be duly appreciated."

Unfortunately, the noble lords and others who had pledged their support, continued to remain anonymous and, if indeed they had the influence which English claimed for them, they did not seem keen to exercise it in the cause of protecting the miners. In fairness, however, it must be remembered that the country's leading social reformers, and more especially Lord Ashley, were much preoccupied at this time in pleading for the four million men, women and children who were working the sort of hours which, Ashley said, the British people would not countenance for a negro in the colonies.

English fussed and fumed at the delay—and the tragic toll continued. At the Landshipping Colliery at Haverfordwest, in Pembrokeshire, more than 40 miners, working on a level under the tidal River Dungleddy, were drowned in a great inrush of water. At Rowbeek Colliery, near Maryport, five workers plunged to their death in a shaft when the winding rope—known to be faulty—broke when bringing them to the surface. The jury awarded a deodand of one shilling to the families, an award which moved the enraged editor of the Journal to declare "...thus the lives of the unfortunate individuals, victims of parsimony and neglect, were estimated at a fraction under twopence-halfpenny! Let all cry shame on this destruction of life, while let it be the object—as it is the duty of all, whether interested or otherwise in works of this nature—to institute a searching inquiry."

After a further series of fatalities, English made another appeal: "If neither the Legislature nor the colliery-owners will take this serious subject into consideration, it surely behoves the latter at least to provide for the widows and orphans of those whose lives are sacrificed. Why not, then, a fund be established for such purpose? There can be no excuse why this is not done, except that it touches the pocket, and if we are to judge by the absence of feeling so generally manifested, we fear that an appeal to the purse will be attended with little avail . . .''

Then, at the end of September, came the most frightful calamity of all. At Haswell Colliery, in the heart of the great Durham coalfield, 95 miners were killed in an underground explosion. This time English decided to go right to the top. If no one else would listen to him, then he must go to the Queen. On October 5, he wrote in the Journal: "We abstain from any comment until the verdict of the jury at the coroner's inquest shall be given, but we cannot allow a week, or even an hour to pass by without again inviting—nay, imploring—our readers to aid the object we have ever had in view, and to which much time and attention has been devoted: that of enforcing some remedial measure by Legislative enactment for the protection of the working collier. Never will anything be done until the strong arm of the law enforces it. Sympathy and interest work not well together, if we may judge from the evidence daily afforded us. In the absence of any effort on the part of others, we have hastily drawn the subjoined petition, which we have reason to believe will be presented to the Queen within 48 hours of the publication of our Journal."

The petition read:

To Her Most Gracious Majesty the Queen.

May it please your Majesty,

The melancholy event of loss of life at Haswell Colliery, Durham, on the 28th September last, by which 95 lives were sacrificed, and their families left in a state of destitution, will, I feel assured, at once ensure your Majesty's sympathy. As a wife and mother, you cannot but feel for the loss sustained by those, who, in humble life, are not only deprived of their partners and protectors, but who are thrown as widows and orphans helpless on the world. Appeals have been made to your Majesty's Ministers, to Parliament, to those possessing wealth, as well as to those whose duty it is to afford protection to the humble instrument by whom their fortunes are created, but without avail.

Let me, then, implore your Majesty, on behalf of the collier and miner, that you will use your Royal prerogative and order a

> *Commission of Inquiry, or institute such other measures as your*
> *Majesty may deem fit, whereby protection may be afforded to the*
> *humble collier and his family. On their behalf, I crave this, and remain,*
> *Your Majesty's humble servant and loyal subject,*

Henry English
Mining Journal Office, 26 Fleet Street, October 4, 1844

And this time it worked. English's personal plea to the Queen as wife and mother, succeeded where official appeals had in the past failed. The barriers were down; this was now a family matter. The Queen talked to her German husband, Prince Albert, who "viewed the subject as one of vital importance and interest". Orders went out from Buckingham Palace and, where there had previously been apathy, there was now bustling activity.

Within a few days, English had on his desk a letter from the Prime Minister, Sir Robert Peel, assuring him that the Secretary of State for the Home Department had taken steps to collect information, and before the month was out the editor was able to tell his readers: "It is with feelings of extreme gratification that we are able to announce, on undoubted authority, that remedial measures will be adopted immediately on the assembly of Parliament. We are now preparing evidence to be submitted to government and to both Houses of Parliament."

And so the wheels were set in motion to provide for the protection of those engaged in the most hazardous occupation in the land. Inevitably they turned slowly, but the terrible tide of unnecessary loss of life was eventually stemmed, and the foundations were laid for the enforcement of a system of regulations which were later to give British mines the reputation of being among the safest and best organized in the world.

Not since the days of Agricola had so much been achieved in the world of mining by one man and his pen.

*

The passion and determination with which Henry English conducted his campaign to reduce fatalities among British miners and to improve working conditions generally, was never allowed to cloud his judgement in other aspects of industrial relations. Quite early issues of the *Mining Journal* reflect the growing attempts by mineworkers to combine and by withdrawing, or threatening to withdraw, their labour, force employers to agree to higher wages or shorter working hours. To the Victorian capitalist such action was little short of criminal and when, in 1844, the miners in practically the whole of the Durham and Northumberland coalfields went on strike—the first major stoppage to be called in Britain—the country was shocked. English himself described the decision as "a blunder", but was quick to point out that "these poor men have put themselves under the guidance of a lawyer ... who had aided and abetted a series of vexatious litigations during the currency of the last bonds and led them to expect that, by standing out, they can enforce from their employers a form of agreement and terms totally at variance with all former custom".

After a meeting of some 20,000 miners, at which they reaffirmed their intention not to return to work until their demands were met; English again emphasised "...it is clear that the poor men arc most grievously deluded by their leaders and unjustly excited against their employers, their wages having been diminished by their own act of limiting their earning to 2s.6d. per day for the purpose of dividing the employment amongst the whole of the Union men."

And, still later, comes a cry which has echoed down the decades of industrial disputes which have followed, and can still be heard with great clarity today whenever the principles of the trade unionist clash with his responsibilities as the family breadwinner: "Out of so great a body there must necessarily be a large proportion of prudent men who, before they involve themselves in certain ruin, would first consider the weighty responsibility which rests upon them as fathers or husbands. Can they, we would ask, endure the children's cry for bread, or the wife's importunities? We feel assured that if these men would exercise the moral influence which every individual possesses within his sphere they would soon find, even among themselves, those who would joyfully co-operate in effecting a reconciliation and bringing the deluded to a sense of duty whereby an equitable arrangement might be entered into, and once again the 'union' would be perfect—that of master and man. Let us hope that such a result is near at hand."

The irony of the 1844 strike situation was that the immediate concern of the strikers was not for the improvement of their own wages, pitifully low though they were, but for their thousands of unemployed workmates for whom there was no form whatsoever of social security or national relief: the lucky ones in employment were willing to accept cuts in their own wages to ensure that their neighbours did not starve.

As one whose interest in mining affairs had stemmed from an involvement in company administration, it was natural for English to be intimately concerned with the development, which began about this time, of the joint stock company and the opportunities for trading in shares, made possible by the establishment of the various mining sections of the London Stock Exchange and other similar organizations abroad. When the *Mining Journal* first appeared most of the Cornish mines were operated on the Cost Book system* and, in the case of the dozen or so important overseas mines, the share units were so large, ranging usually between £25 and £100, that transactions were few and not easy to negotiate. It was not in fact until well into the third quarter of the 19th century that investment in mining really became a practical proposition for the general public, the growth of joint stock activity being spurred first, in Britain, by the gold discoveries in Australia, South Africa and Canada, and later by the beginnings of the big base metal developments which continued into the present century.

In his 20 years as editor, English wrote much more about company affairs than he did on mine accidents, industrial relations, or any other single subject. At one time he was persuaded by friends to set himself up as "agent in the purchase and disposal of mineral and other property, as also that of stock and share broker", pointing out to readers of the *Mining Journal* that "the extended circle, arising from his immediate connection with that publication, may be fairly pre-

---

* Under the Cost Book system shareholders were usually entitled to withdraw their investment at any time provided that they paid their proportion of existing net liabilities up to the time of withdrawal.

sumed to afford more than ordinary facilities and advantages in the transaction of business as mineral estates and mine agent, to which may be added an intimate knowledge for the past 15 years of the mines and collieries of Great Britain and Ireland, as well as some parts of the continent".

History does not record whether or not English's "moonlighting" as a broker prospered—or how he reconciled these business interests with the impartiality necessary for an editor dealing with the financial affairs of the mining companies. However, neither these, nor any other, interests took up enough time to prevent him from embarking on a campaign which he pursued, if not with the same passion, certainly with no less persistence than he had devoted to his efforts for mine safety.

When the *Mining Journal* first appeared in 1835 there were virtually no facilities in Britain to study either the science, or the art, of mining. In fact, there was little opportunity to study any branch of science at all. The universities were still under the antiquated dominion of classical teaching and the little scientific instruction that was given was confined to the lecture hall; laboratory work for students was unknown. And, of all the sciences, geology was perhaps the most benighted, still the subject of great controversy between the Neptunist disciples of Werner, and the Plutonist followers of Hutton, who waged bitter war about its fundamentals.

One thing, however, that Britain did have, was a geological map. Some 50 years earlier, William Smith, a surveyor by profession, a plain man of little outward culture and no wealth to help him undertake great things, recognized as he went about his daily work, that the rock formations over which he passed bore definite relationship to each other. All over England he tramped, surveying and taking notes: in 1790 he published his *Tabular View of the British Strata* and finally, in 1815, he completed his map. An indifferent government rewarded him with a pension of £100 a year; today he is the acknowledged "father of English geology".

At about the same time as "Strata" Smith was working so industriously and unobtrusively at his geological researches, a young boy with an enquiring mind, was rambling over the cliffs which fringe the sea at the Devon village of Charmouth, where he lived with his widowed mother. Henry Thomas De La Beche was fascinated with the shells and fossils embedded in the cliffs and was determined to discover how they got there. Sent to military school when he was 14, he soon tired of soldiering in peacetime and left the army at the first opportunity to return to his first love, geology. He studied mineralogy and, following the footsteps of William Smith, he began, with the encouragement of the Board of Ordnance, to prepare for a fresh geological survey of Britain. Eventually his work was recognized by government and the Geological Survey of the British Isles was launched.

By the time Henry English was writing his first editorial for the *Mining Journal*, De La Beche had collected a mass of notes, plans, surveys, drawings, fossils and minerals of such obvious importance to the country's mining interests that he sought means of bringing his work to public notice. In many of the most potentially productive mining areas, the ignorance of mine-owners, management and men was hampering development. Time and energy were being squandered; shafts were being sunk and plant erected in places where there was not the slightest hope of a return, whilst rich mineral fields were ignored. At Freiberg, Werner had taught the necessity of keeping geological and mining records, but in Britain the lesson still had to be learned.

The *Mining Journal* was a ready supporter of a move to set up a national mining records office, coupling with it the need for a school to be devoted to teaching mining, both in theory and in practice. The Journal had been in production for only a few months when English wrote: "Indebted as this country is in a great measure to its metallic products for its national wealth, it has ever been a matter of surprise and regret that we should be without any school of mines or, indeed, *any* medium whereby information may be obtained by the miner, while Germany and Hungary boast of their colleges at Freiberg and Schemnitz, which are often visited by those interested in the mines of this country who possess the pecuniary means of obtaining the information they desire." On the specific issue of a records office, English declared: "The importance of a central office, from which may be obtained information connected with our mining interests, is too obvious to require advocacy on our part."

But the editor vigorously continued such advocacy, repeatedly drawing attention to this great want. Meanwhile, De La Beche, with the help of Thomas Sopwith, well known in the mining world (and father of the founder of the Sopwith Aviation Company), induced the British Association to back his cause with government, with the result that a small museum and records office was set up in London near Charing Cross.

By the 1840s, the Geological Survey had become an important government department, having been prised apart from the militarism of the Ordnance Survey.* However, it still fell far short of De La Beche's ambition to establish larger and more permanent premises and to set up a school that would include the practical teaching of metallurgical processes and the rudiments of mining. By now a more favourable atmosphere was beginning to develop for the furtherance of scientific work, stimulated by the interest of the Prince Consort who, anxious to promote the cause of education and scientific endeavour, had conceived the idea of a gigantic international exhibition in London. At this opportune moment, De La Beche—now Sir Henry—pushed his claim for a more fitting home for his ever-growing collections and records, and in 1848 government approval was given for building to start on a Museum of Practical Geology, in Jermyn Street.

From then it was but a short step to the ultimate goal. The pressure that Sir Henry was able to bring to bear was irresistible. Memorials from the chief mining centres and manufacturing towns, emphasizing the importance of establishing provincial schools of science in conjunction with a central metropolitan school, were sent to the Royal Commissioners of the Exhibition of 1851. A House of Lords committee backed the proposal; so did the leading scientists of the land; the mineowners, and the mineworkers, gave their enthusiastic support. And,

---

* The uniform of the Ordnance Survey was a tight-fitting frock coat with gilt buttons, embossed with the royal crown and crossed geological hammers, with the letters O.G.S. Long after the uniform had been discarded the buttons were retained as an emblem, surviving today as the badge of the Royal School of Mines.

behind all, was the strong influence of Prince Albert who, in the midst of the preparations for the Great Exhibition, had not forgotton Henry English's impassioned appeal for the safety of British miners a few years earlier.

And so the Royal School of Mines came into being—initially under the cumbersome title of the "Government School of Mines and of Sciences applied to the Arts"—housed, with the Geological Survey and the Mining Record Office, at the Museum of Practical Geology. Both the new Museum and the School were opened in 1851, the former by Prince Albert in May, less than a fortnight after the inauguration of the Great Exhibition, and the latter in November, a week or two after the last of the seven million visitors had passed through the exhibition gates at Hyde Park. Each contributed to the further advancement of scientific knowledge, the profits of the venture being devoted to new institutions in Kensington, where the Royal School of Mines is today located—appropriately, in Prince Consort Road.

Whilst not so intimately concerned in its promotion, the *Mining Journal* also gave enthusiastic support to the development of mining education in the West of England. Here, the Mining Association of Cornwall and Devon took the initiative by organizing evening classes in various centres to help the largely untaught miners. In the course of time, more ambitious courses were introduced and schools were established at Redruth, Camborne and Penzance, in the heart of the copper and tin producing districts.

Early in the present century these three schools were amalgamated to form the present School of Metalliferous Mining, better known as the Camborne School of Mines, one of the Commonwealth's premier colleges, whose students have over the years made a significant contribution to the development of world mining.

CHAPTER FOUR

America's Golden Age

An early mining reporter in Arizona

Save for coal, Britain had reached its peak as a mining country by the middle of the 19th century. Metal production in other parts of Europe was also declining, or failing to keep pace with expanding needs, and industrialists were beginning to look around, somewhat anxiously, for new sources of mineral wealth.

The discovery of gold in California in 1848 and in Nevada a few years later, brought the North American potential sharply into focus. It gave rise to the biggest treasure hunt the world had ever known and although, for most of those who took part, it ended in disappointment and for some, disaster, it opened up the West in rapid and sensational style, providing—as mining has so often done in history—the trigger for much wider economic and industrial expansion than the mere panning of gold from mountain streams.

The American colonists had always been interested in mining. Indeed, one of the declared purposes of establishing the first settlement in Virginia was to prospect for minerals, but precious metals, such as the Spaniards had discovered in Central and South America, were not to be found. The only mining activity that developed widely from colonial times was iron-making in smelting works scattered along the eastern seaboard, though early production was based more on the plentiful supplies of wood for charcoal than the availability of a high-grade ore, most of the deposits being of bog-iron. There is also some evidence of lead mining, particularly during the War of Independence, though most of the lead used for the American Army's bullets was secondary material. It is on record that soon after the war began General Winthropp, of Connecticut, ordered a leaden equestrian statue of George III (whose obstinacy was the main cause of Britain losing her colonies) to be melted down to produce 42,000 bullets.

Interest in copper began with the investigations along the shores of Lake Superior by Douglass Houghton, whose reports in the 1840s convinced the world of the vast reserves in this region at a time when copper resources in Europe were beginning to dwindle. Serious mining was begun when the Pittsburg and Boston Company was organized in 1844 to operate the Cliff Mine in the Eagle River district. Here, the fissure veins contained native silver as well as copper and an adit, driven into the base of the cliff, ran into a great lump of metallic copper. The discovery was important for the future of copper exploration, showing that mineral-bearing boulders scattered erratically on surface can have their origin in lodes below ground.

Some 50 years earlier, at about the same time as "Strata" Smith was tramping the English shires to prepare his geological map, a Scotsman, William Maclure, was similarly engaged in North America. He surveyed from the St. Lawrence River to New Mexico, crossing the trackless Allegheny Mountains alone no fewer than 50 times, and in 1809 he published his observations, together with a geological map (predating Smith's) showing a systematic classification of rocks. His work was of immense value to future explorers for minerals, and, indeed, if William Smith is to be accepted as "father of English geology", then Maclure must be given similar title in the American context.

However, Maclure did not penetrate into California and it was the sharp eye of James W. Marshall that spotted the glitter of gold in the millrace he built in the Coloma Valley and which resulted in the first great rush for riches in 1848. Within a year, 80,000 people from all over the world had reached the West Coast, by caravan and wagon train from across the dusty plains and rocky mountains, or by ship via Cape Horn, a voyage which took between six and eight months. In a single month, 60 ships sailed from New York, and 70 each from Philadelphia and Boston, as well as hundreds from many other ports; in one day, 45 ships arrived in San Francisco with fortune-seekers from many countries.

Fortunately for the first prospectors, few of whom had any technical knowledge of mining, the geological evolution of California had brought about conditions favourable to the simplest methods of recovery. In early geological time, a subterranean intrusion of molten material had produced countless small veins of gold-bearing quartz that made up the so-called Mother Lode. Eventually, the overlying rock eroded, exposing the veins to westward-flowing streams that washed away the surrounding quartz and concentrated the gold in sand bars and along the beds of the streams, resulting in one of the richest gold-placer regions ever known. Little or no knowledge of mining was required, the "forty-niners" working with little more than shovel and pan, digging out the gravel from the stream-beds and washing it in their pans until—if they were lucky—only a glittering streak of gold particles remained.

Reliable records of the early operations are scanty, largely because most of San Francisco's newspapermen and printers had joined all others in the mad rush to the "diggings", leaving a ghost town behind them. Later, as sanity returned, the picture became clearer. Slowly, with an increasing element of capital investment, more sophisticated techniques began to be employed; river mining, by which streams were dammed and diverted so that gold could be dug from the dry beds; hydraulic mining, which meant washing away the gold-bearing gravel with a high-pressure jet of water; and quartz mining, or underground mining of the gold-bearing veins.

Underground mining accounted for about 10 per cent of California's gold production in the 1860s and 30 per cent in the 1870s, the techniques being learned from veterans of the world's older mining regions—Cornishmen, Georgians, Spanish Americans and Germans. Since the early shafts were quite shallow and the underground workings small, miners were spared the more difficult problems of hoisting, timbering and ventilation. Even so, it took several years of trial and error to learn how to sink a shaft, break the ore and hoist it to the surface. Since the gold was not free, unlike in the placer deposits, the quartz had to be finely crushed to release the metal particles. For this operation, much improvisation was necessary until many practical improvements resulted in the Californian stamp mill, which was to become one of the State's contributions to mineral technology.

In the wake of the California gold rush, the fabulous Comstock Lode was discovered in 1859 in the Washoe Mountains, a range of sagebrush-covered hills that jut eastwards from the Sierras into the Great Basin. About 2,000 feet below the summit of Mount Davidson, the most easterly hill, was a vein of decomposed quartz that was to yield gold and silver worth $30 million during the next two decades. Most of those who swarmed into this area were refugees from the decaying placer-camps in California, unsuccessful prospectors, or inexperienced hands from the Missouri frontier, along with the usual complement of speculators, saloon-keepers, merchants, outlaws and drifters.

The main lode was discovered by James Fennimore, popularly known as "Old Virginny", a bibulous prospector, who gave his name to Virginia City, which grew from a group of shanties which sprawled over the breast of the mountain. It was here that *Territorial Enterprise* established itself under its daredevil editor, Joe Goodman, to wage constant battle with its rival, *Virginia Daily Union*. The thirst for news among the mining community was intense as America now stood on the brink of civil war. The Pony Express left St. Joseph, Missouri, on the afternoon of November 7, 1860 with despatches of Lincoln's election and arrived in Virginia City in the early hours of the morning on November 14, 1,120 miles covered at an average of about 7 miles per hour. In San Francisco, the leading newspaper reported the event beneath the flying headline, "Only Seven Days from the East—New York State goes fifty-thousand for Lincoln". With Lincoln's election, the feeling between North and South was stretched taught. Within a few weeks, South Carolina headed the secession of the Southern States, and in the following April the guns of the Confederates fired on Fort Sumner to begin four years of bitter war.

The Comstock was sufficiently isolated to be relatively unaffected by the conflict. Its inhabitants had plenty to worry about in their own small world, huddled against the mountain. They also had plenty to laugh about for the *Enterprise* had hired a young man with rare talent. His name was Sam Clemens and he set out to amuse them with whimsy and rhetoric, aided and abetted by the paper's star reporter, Dan de Quille, perpetrator of hoaxes and fabricated stories, and who was later to write the first authentic story of the Comstock.*

The miners put cap and bells on Sam's head; their encouragement had more to do with moulding him into a humorist than any other influence. Sam caught the hilarious mood and, with tall yarns and literary burlesque, he played to the rough and rowdy gallery of miners and soon achieved a reputation far beyond the confines of the Forty-mile Desert. It was here that Sam Clemens became Mark Twain, and he stayed for two rollicking years before one of his escapades went a little too far: a warrant was put out for his arrest for duelling, and he slipped quietly out of Virginia City under cover of darkness in an east-bound stage, and into a bigger world in which he gave to *belles lettres* a name that is still among the best known in American literature.

*

This, then, was the ground on which the seeds of American mining journalism were sown in the 1860s. The activities in California and in the Washoe Mountains, crude though they may have been in their early stages, also represent the beginning of the modern history of mining in the United States which has developed into the complex industry that it is today.

Early experience had shown that major development depended on professional expertise, capital and equipment, and these were the interests to which the first mining papers set out to appeal. The completion of the transcontinental railway system after the Civil War was a further factor in the consolidation of the industry from a host of independent mines, mills and smelters into integrated operations run by large corporations. Before the Sixties were out it was possible to travel from San Francisco to New York by train. And in these two cities the first mining journals of importance were established.

**History of the Big Bonanza.* The author's real name was William Wright, but he seems rarely to have used it.

Appropriately, the first was in San Francisco, which by 1860 had largely recovered from the desertion of its inhabitants to the Mother Lode country—and from the difficulties of rehabilitation when so many of them returned empty-handed and disappointed. The bay had been largely cleared of the abandoned ships which choked it during the great influx of fortune-seekers; merchants were extending wharves to deeper water; and, on shore, shacks and tents, which had multiplied amidst mire and dust, were cleared away as more permanent residents set up homes and offices to provide equipment, stores and capital for the miners.

On May 24, 1860, from its offices at 172 Washington Street there was issued the first number of the *Scientific Press*, a "journal of science, art, manufactures, chemistry and inventions." There was, curiously enough, no mention of mining. Its proprietors and publishers were George H. Winslow and Company, and its prospectus announced that its columns would be devoted to "the interest and advancement of the Mechanic and Working Man—to instruct them in all that is useful and advantageous to their trade and profession, and to inform them of all the latest and most approved discoveries in science and art".

And that was not all. The prospectus went on to proclaim that "inventors of useful machinery and instruments will find this the only medium in the State wherein their claims for patents may be successfully prosecuted. It is therefore requested that those interested should leave with us their plans, drawings and models with explicit specifications and such other information as they may possess relating thereto."

Whether anyone took advantage of this offer and left their precious plans in the little office over a store is not known, but any local inventor who may have been tempted to do so would have had his confidence a little shaken a couple of issues later when it was announced that Mr. Winslow would in future have nothing to do with the enterprise, but his former partner, Mr. J. Silversmith would continue to publish the paper which was to be edited by "an association of scientific men". In September the same year it was decided to publish daily during a fair in the Mechanics Pavilion that month, but of greater significance was the announcement that, owing to their growing importance, the mining interests of the State would thereafter be the chief feature of the paper.

Silversmith was succeeded as publisher in 1862 by W.B. Ewer and C.W.M. Smith, who were joined the following year by Alfred T. Dewey, the company then becoming Dewey and Co. Alfred Dewey had been in the newspaper business for some time, in Westfield and Springfield, Massachusetts, and more recently at La Porte, in Sierra county, California, where for five years he had published the *Mountain Messenger*. He changed the name of the paper to *Mining and Scientific Press* and directed the affairs of the company for the next 30 years.

At first, the new owners appeared to be of a migratory nature; possibly they found it cheaper to move than to pay rent, and it must have been extremely unsettling for the staff to have to change office eight times in the first five years. However, by the end of the decade some stability had been achieved—not only at *M and SP*, but in San Francisco generally—and when, in 1869, Almon D. Hodges joined the paper as a junior editor, he arrived in the city in "an age of simple, comfortable living, hopeful happiness and unstinted hospitality".

Forty years later, when *M and SP* celebrated its 50th anniversary, Hodges set down some of his recollections of California's golden age at the time when he joined the staff. The mines, he said, were then yielding annual dividends of 25 per cent or more; at least, when a mine was offered for sale, if the examining engineer could not report that it would probably return the purchase price in four years, his clients refused to buy. This was the general rule among those who could be seen any week-day on Pine Street, dabbling in stocks and ready for a good mining investment. Most of them had worked in the goldfields, where they had accumulated sufficient for their wants and with 25 per cent gain on most of their dealings, they soon had enough for more than the simple life.

"Many of them", recalled Hodges, "had pleasant homes across the bay, and it was their custom to invite thither, to spend Sunday, half a dozen to a dozen guests, who were then regarded as members of the family and given the freedom of the place. Did you desire to give a ride to a charming young lady whom you, perhaps, had met for the first time, it was merely necessary to order a hostler to harness up a gig. Everything on the place was at your disposal. With these men it was 'easy come, easy go', and they generally spent all their income on free-handed hospitality. They lasted until the days of the Bonanza stock speculation which wiped them out."

San Francisco was at this time a materialistic city, quite unattuned to literature and the arts. It was told of one rich man, who was building a large and expensive residence, that he and his wife declined to purchase some genuine "old masters" because they "did not care to have any second-hand pictures in their house". There were plenty of bright young artists around who, failing to sell their pictures, contributed articles on all subjects to the papers whenever they could get them accepted, thereby making life difficult for established journalists. Young Hodges got $16 a week from *M and SP*, which was enough for him to live decently and put some money in the savings bank. When he eventually received $150 his income exceeded that of all but a few of his contemporaries. "But," he reflected, "it was a bright and jolly brotherhood, a genuinely Bohemian brotherhood of the pen and the brush."

Life on the *M and SP* was not, however, without its complications. Alfred Dewey had an ambition to become a big publisher and accordingly launched, or bought, a large number of ephemeral papers, invariably crippling the resources of his chief journal. He was also prone to make constant changes in the make-up and even the names of his papers, and he clung tenaciously to certain features which the most conservative of his staff considered inappropriate to a technical journal struggling to achieve professional standards. Among the papers he owned was a *Ladies' Home Journal* and he constantly emphasized its usefulness to the *M and SP* as being a means by which awkward spaces could be filled by cooking hints and recipes important to bachelor miners.

Much greater embarrassment was to result from Dewey's purchase in 1871 of the *Pacific Rural Press*, the leading agricultural paper on the Coast. Ewer, who had edited *M and SP* since the take-over from Silversmith, had been in journalism since his youth and had edited papers in both the East and West. He had lived in the mining regions and well understood conditions on the Coast. But he had always had a consuming interest in farming; the company's latest acquisition gave him the opportunity he had always sought, to be editor of an agricultural journal, Editorship of *M and SP* passed to Charles G. Yale, who had been recruited only a few months earlier with no experience in journalism. Yale had set out to become a mining engineer, but when his father* died he had to discontinue his studies and look for a job. Dewey decided that he fitted the company's requirement for a "young man, who was college-bred and knew something about mining" to act as assistant editor. Poor Yale was therefore "thrown in at the deep end" when Ewer unloaded the responsibilities of *M and SP* onto his shoulders in order to concentrate his own efforts on the farming paper.

Unfortunately, at about this time the interests of mining and farming were thrown into sharp conflict by the devastating effects which hydraulic mining was having on the Californian countryside. In areas where there were extensive deposits of deep gravels, hydraulic mining had become an effective and very inexpensive mining method, gold being recovered at the cost of a few cents per cubic yard of gravel. But the more effective this method became, the more destructive it was to the landscape. Whole hills were washed away, leaving huge open cuts in the previously forested mountainsides. At the end of the sluices, the dirt, gravel and rocks spread out in a filthy mass until heavy winter rains swept it down into the lower valleys at the foot of the Sierras. As much of this rich farmland had been brought into cultivation, the frustrated farmers sought injunctions to restrain the miners from ruining their crops.

Dewey, as publisher both of the country's only mining paper and the leading West Coast farming journal, was in a dilemma. His personal sympathies were undoubtedly with the farmers, but he was reluctant to take either side, for support for the one would undoubtedly injure the paper which catered for the interests of the other. So he decided to sit firmly on the fence—as a result of which *both* papers suffered loss of prestige and, on occasions, aroused much public fury.

---

*Gregory Yale, whose treatise, "Mining Claims and Water Rights", was a pioneer work on American mining law.

"Neither paper", wrote Yale many years later, "had much to say on the subject and, indeed, ignored it as much as possible. I was given strict orders by the head of the firm to write nothing about hydraulic mining, to put in the paper only 'official news', and to show him any items which were to appear. Here was an unpleasant situation for an editor: to be muzzled on the very topic most interesting to the community which his paper was supposed to represent. Many a savage letter did I get, and many a personal growl from prominent mining men for the course I was taking. For hundreds of these men knew me personally and did not know either Mr. Ewer or Mr. Dewey. The latter dominated the firm and its policies, Mr. Ewer being something of an invertebrate, with no opinions opposite to those of Mr. Dewey or, at least, none that he advanced. He was a kindly old gentleman and an excellent associate, but he often aroused my anger by his way of dodging questions and of taking no side, nor giving any opinion."

The negative position taken by *M and SP* doubtless injured the paper materially for the farmers eventually succeeded in their suits, and the suppression of hydraulic mining resulted in the closing of many such mines and the de-population of a large number of small camps and communities. It was not until the paper was sold some 20 years later to Alfred Holman and J.F. Halloran that its editor was allowed to write freely on hydraulic mining.

Charles Yale was still in the editorial chair when this change of ownership occurred and he at once set about repairing the damage, not only to the paper but to the mining community which had suffered from the injunctions. At a convention called by the miners of Placer county, at Auburn, he was among those who drew up an appeal to the people of California for assistance. He was also appointed a committee of one to try to organize the movement in San Francisco and to get the daily press in line to help the miners again.

On his return to the city, Yale, with the assistance of William C. Ralston, organized the San Francisco Mining Association, of which he became the permanent secretary, and he also persuaded the daily papers to accept a series of articles which he wrote on the condition of the hydraulic mining industry. The result of these efforts was the foundation of the California Miners' Association. At the first convention of that body, Yale, as editor of *M and SP*, was appointed to represent the miners, and Robert Devlin to represent the anti-debris men, or the conservationists, and they together drew up the memorial to Congress, which resulted in the passing of the Caminetti Law permitting hydraulic mining under certain conditions.

Despite the time that *M and SP* took to clarify its public attitude to the hydraulic mining issue, attempts to dislodge the paper from its position as the accepted organ of mining in the West, all met with failure. At various times more than a dozen papers were started in opposition in San Francisco, but none of them lasted very long. Many people thought that *M and SP* was too old fashioned and conservative and that its proprietors had no enterprise. To some extent this was true, but the paper was too firmly based in the mining communities to be easily displaced. In San Francisco, its editorial offices became a meeting place for the leading personalities of the industry, and some of them became regular contributors to its columns. Among them was Dr. Henry De Groot, pioneer miner and the first newpaperman (for the *New York Herald*) to interview James Marshall and others concerned with the discoveries at Coloma. Other names in the early by-lines included those of Guido Küstel, C.H. Aaron, C.A. Stetefeldt, Almarin B. Paul, George Davidson, J.S. Phillips, Melville Atwood and Henry G. Hanks, all of whom were closely associated with the development of mining in California.

Dr. Rossiter Raymond, who was to become dean of American mining writers, and, chief administrator for nearly 30 years of the American Institute of Mining Engineers, which he helped to form, recognized *M and SP* as serving two great ends of equal value. These he described as "the encouragement of suggestions, even from the ignorant, which might contain facts of importance; and the enforcement of the principles of science, under which alone such suggestions can be suitably weighed and utilized". Raymond believed that science could not afford to despise the observations and experience of those "practical" men who were not scientific. "Such men", he declared, "should have a hearing. In many instances they have something to say that is worth notice. And a journal which records their statements and views is furnishing material for the science of the future."

Increasingly, *M and SP* did just that, keeping a fine balance between straight reporting of the progress being made at the mining properties and tempering the practical outlook of the miners with the views of those with more technical and scientific background whom it could persuade to write for its pages.

In 1898, Holman sold his interest in the paper to Halloran, a man of considerable previous experience in newspaper publishing, and he and Ewer, now getting on in years and with little or no editorial influence, continued to own most of the stock until it was bought by T.A. Rickard, a mining engineer with a distinguished record internationally, who had lately turned to journalism. As Rickard and Halloran shook hands on the deal at the company's office in San Francisco, the building trembled with the shock of a minor earthquake. "That always happens when *M and SP* is transferred," quipped Halloran. A few months later the office and plant were in ruins—as was most of the city, laid waste by one of the most disastrous earthquakes of all time.

A new *M and SP* emerged from the devastation under Rickard's energetic direction. One of the first things he did was to lay the ghost that had haunted so much of its past: he sold the *Pacific Rural Press*. And with the proceeds he bought a linotype machine. The old type cases, from which the paper had previously been set by hand, were swept away with the rest of the earthquake rubble, and the venture which nearly ruined the company in the 1870s enabled *M and SP* to make a fresh start in the 20th century with the most modern equipment.

CHAPTER FIVE

Growing up in New York

Richard P. Rothwell, editor of high courage and high principles

As the cradle of large-scale mining in the United States, the West had naturally become the first important centre of mining journalism. It was not long, however, before it began to be recognized that ultimately New York was bound to play an increasingly important part in the industry's development. It was here that most of the capital would originate; the purchase of the bigger and better machinery required for the expanding properties would be arranged here; the headquarters of many of the large mining companies would need to be at the centre of the nation's commerce and communications; and most of the products from the mines would flow back to the East for marketing and distribution.

Among those who were quick to anticipate these changes, which would make New York the hub of the mining world, was a young man named Charles Callis Western. Born in New York in 1842, he had gone to California as a boy of 16; he ruined his health working with pick and shovel at the gold diggings and when he had saved enough money he moved on to San Francisco to start an illustrated paper. When that failed he joined the many other struggling youngsters in the city, supporting himself as an artist and wood-engraver. Doubtless he saw the beginnings of the *Mining and Scientific Press*; at any rate, when he returned to New York in 1861 he had already conceived the idea of starting a mining paper in the East and, with his elder brother Benjamin, he formed the publishing firm of Western and Company for this purpose. It must have been extremely difficult trying to begin an enterprise of this sort in the midst of civil war and it was nearly a year after the defeat of the Confederate Army that the first issue of the weekly *American Journal of Mining* appeared on March 31, 1866.

Its editor, George Francis Dawson, introduced himself in a salutatory, in which he declared an abhorrence of "puffery", an intention "utterly to ignore politics" and a determination, "as far as lies within my capacity", to make the journal a success. Very little is known of Dawson's background but a glance at the first issue indicates that he was a journalist with a natural flair. Of special interest is an editorial which gives a thumbnail sketch of the state of the industry at that time:

"The gold and silver mines of Nevada, Montana, Idaho, Colorado, Arizona, Oregon and New Mexico, never yielded larger returns. Those of California manifest decided symptoms of improvement. The product of the great copper region bordering Lake Superior exhibits no decrease; and the coal, iron and lead mines of Pennsylvania, Wisconsin, Illinois, and other States, certainly hold their own. Most encouraging accounts as to the mineral wealth of Virginia, Georgia, and others of our Southern States, also continue to come in, to the great astonishment of those who, in their eager westward gaze, have hitherto overlooked the vast treasures lying buried almost at their own doors. The petroleum product, in spite of the heavy tax imposed upon 'crude', is growing, and when Congress shall have lifted the galling burthen from the producers' shoulders, it must become a highly remunerative branch of the mineral industry.

"The excitement relative to Montana and Idaho does not seem to have been materially affected by the glowing accounts from New Granada; and the more recent reports from the Big Bend mines in British Columbia, although bearing the imprint of truth, serve but to increase the bent of many adventurous minds to swell the populations of those two favoured Territories. The coming month of May will undoubtedly witness an unusually large migration to the far West of those anxious to make their fortunes by practical mining . . . Colorado, doubtless, will also receive a fair share of the overland pilgrims into her bosom . . . as Spring advances confidence in mining investments will also advance; but after the many severe lessons they have learned, it is to be hoped that capitalists—whether large or small—will not again embark rashly on wild ventures. The insane spirit of speculation has been too common; mining can never be conducted unless as a legitimate business but, as a business, it can, without doubt, be carried on most remuneratively."

Already, in its first issue, the New York-based journal was taking a much broader view of things than San Francisco ever could under the prevailing conditions. Dawson's reference to petroleum was backed by an original paper on the subject by Francis E. Engelhardt, Professor of Chemistry at St. Francis Xavier's College, whilst Dr. E.H. Grant, State geologist of Virginia, expanded on the mineral potential of that territory. An article on compressed air prophesied that it would soon come into general use underground and surmised that the era of mining machinery was beginning, adding that this could only be achieved by ignoring the "contracted ideas and clannish habits of Cornish mine labourers—perhaps, of all men, the most obstinate and troublesome to deal with".

A New York metal market was already flourishing and its reports show that Lake copper was then selling at 30–31 cents per lb (on the London Metal Exchange, "best selected" copper was just under £100 per ton), American iron was at $45 to $47 per ton, and best cast-steel 22 cents per lb.

It seems to have been customary in the early issues to describe and illustrate one mining machine in each number. But the machinery firms were slow to reciprocate by advertising, and it is curious that the first advertisements in the paper were of patent medicines. However, a growing list of editorial contributors included some noted names: Professors Henry Wurtz, of New York; H. Dussauce, of the French Polytechnic School; Paul C. Morton, of Oglethorpe University; and J. Van Cleve Phillips. A complete coal trade report, compiled especially for the Journal, covering from 1830 to date, appeared in the issue of January 26, 1867, and is undoubtedly the best record of the country's early development in coal production. In this year, too, was reviewed the first Government report on the Western mining industry—a preliminary review by J. Ross Browne, Commissioner of Mining

Statistics, which was criticised with a severity not wholly deserved, the little book apparently being judged by a standard far higher than its pretensions.

Both the Western brothers played active parts in producing the Journal in its early years. Charles took a hand in the scissors-and-paste operations by which the "Mining Summary" (which was soon carrying news from overseas, including Australia) was compiled, and Benjamin would gather in the advertising, having, as one member of the staff later recalled, "a very pretty knack of making cheques do double duty when occasion required".

In the decade in which the Journal was born, the struggle in the Western prairies and mountains between the white settlers and the Indian tribes, who resisted their encroachment, was at its most intense. Despatches from the West came laden with reports of atrocities by the Redskins, and the Journal frequently added its voice to the clamour for reprisals. The paper carried a report of the Fort Laramie massacre in 1866, one of the consequences of which was that miners in several of the Western Territories were obliged to go prospecting in armed bands. On another occasion the Journal denounced Government Commissioners for "bribing Indians to sign treaties with presents of firearms and ammunition with which they are enabled to shoot down miners, rancheros and travellers . . . knowing as we do, the treacherous, crafty, cruel treaty-despising character of these degraded aborigines, we lift our voice against such methods of conciliation." A couple of years later, however, the Journal was warning its readers, not about rifles, but of the deadly effect of Indian arrows, quoting a report that at Laramie only two of the 82 soldiers massacred were killed by bullets—and they probably shot themselves after being wounded by arrows, "to avoid falling alive into the hands of their cruel foes".

On scientific subjects, the Journal noted the efforts of a Mr. Wenham to fly the length of the Crystal Palace during a meeting of the British Aeronautical Society; a demonstration of a new apparatus called the telephone (intended to "transmit a musical melody by telegraph"); the discovery of diamonds in South Africa, and of borax at Clear Lake, California. The obituary columns included notices of the deaths of Colonel Drake, the first man to strike oil in Pennsylvania, and of

James Marshall, discoverer of gold in California; both men, it was recorded regretfully, died paupers.

Early in the Journal's second year of publication two names began to appear among its correspondents for the first time: Rossiter W. Raymond and Richard P. Rothwell. Together, these two men were to dominate the affairs of the Journal for the rest of the century and to lay the foundations on which America's leading mining publication has since been built.

Already in the Sixties, both men—Rothwell more particularly—had achieved reputations in the field of engineering. Rothwell came from Ontario, Canada; Raymond from Cincinnati. Both had studied at the Mining Academy at Freiberg in Germany. After some experience in England, France and Canada, Rothwell went to the anthracite-coal fields of Pennsylvania, where he worked for 10 years, designing and building engineering equipment, making topographical and railway surveys and doing much other professional work. Raymond, after three years in the Federal Army during the war, had gone into partnership in New York with Dr. Justus Adelburg, a well-known chemist and technologist with considerable practical knowledge of mining on the West coast, and it was not long before the firm of Adelburg and Raymond were doing a bustling business as consultant mining engineers.

It was during his association with Adelburg that Raymond was asked by Western and Company to contribute to the Journal—first, anonymous news paragraphs; then signed articles; and, finally, he was invited to become editor when Dawson resigned after little more than a year in the chair. The job was not to be a full-time one but, even so, Raymond was a little afraid that it might prevent him from putting his professional skills as a mining engineer into practice. In fact, the effect was exactly the opposite. A series of articles which he wrote on the relationship of the Federal Government to the mining industry attracted the attention of some senators in the Pacific States and when, in 1868, Ross Browne was suddenly appointed U.S. Minister to China, Raymond was invited to Washington and offered the job of U.S. Commissioner of Mining Statistics. He was then only 28. Raymond always felt that such an honour and opportunity would never have come his way but for his association with the Journal and, indeed, the paper was to remain a major source of the professional knowledge and influence that he was to achieve. It is a remarkable fact that in the next 40 years not a single issue of the Journal appeared which did not include a contribution from his pen—and he continued writing for it for many years afterwards—despite the fact that he severed official connection with the paper in 1889.

It was Raymond who, in 1868, changed the name of the paper to *The Engineering and Mining Journal* (to be widely known ever since by its contracted title, *E/MJ*), explaining in an editorial that the reason was "to express more accurately the comprehensive character which we have attempted to impress upon the Journal, but which the former title has, in many cases, covered from recognition, so that numerous parties, interested in the cognate branches of engineering, have passed us by, supposing that we devoted ourselves exclusively to mining".

It was a timely change when it is remembered that in these days the lines were much less sharply drawn between the different branches of engineering than they are today. Nearly every piece of engineering equipment which was developed or improved in the second part of the 19th century was described and illustrated in the Journal and, like the contemporary *Mining Journal* in London, it could not help but be concerned with the enormous railway development which was helping to shape the future of both countries. Moreover, mining at this time was becoming more and more of an engineering business rather than a free-for-all scramble for riches.

Between 1869 and 1871 the Journal ran a series of important articles on the copper deposits of the Keweenaw Peninsula, by Dr. Herman Credner, of Leipzig, translated by Max Müller; and there was special correspondence from Professor F. Stapff, of Sweden; and from Dr. T. Sterry Hunt and James Douglas, who were then working on their process for the treatment of copper materials which was to lead to the establishment of the first copper electrolytic refining plant at Phoenixville, Pennsylvania.

But the name that was commanding most attention among the technically-interested readers of the Journal was Rothwell. A series of articles on the economy and efficiency of various mine ventilation systems and equipment was followed by detailed descriptions of methods of dealing with fire-damp and underground fires. Rothwell's mastery of this subject had derived not merely from theoretical study, but from hard, practical experience, in the course of which he had more than once displayed great heroism. His articles gave no hint of this and simply recorded his professional conclusions, but later—much later—some accounts of his personal adventures began to emerge.

Once, at Pine Ridge, a particularly fiery colliery, Rothwell led his men to clear a shaft which was rapidly filling with fire-damp and succeeded in preventing a major explosion with only seconds to spare. In a disaster at West Pittson, where 24 lives were lost, he organized parties to rescue the entombed miners. At first, only dead bodies were found, but Rothwell's knowledge of the underground workings and of the mine's ventilation system made him sure that many other men who were trapped could be reached and saved. He led volunteers to break through into a small chamber, into which some 50 men had crawled and walled themselves up against the gases. In successive trips through the stifling atmosphere they were brought out, some living, some dead. The rescuers escaped death though some, including Rothwell, were revived only after much difficulty.

Rothwell's articles soon led to an invitation to become an associate editor of the Journal and when, in 1871, the Western brothers, who were by now interested in several other publications, decided to sell the mining paper, Raymond, Rothwell and Willard P. Ward (another associate editor) founded the Scientific Publishing Company in order to buy it.

Unfortunately, whilst the paper itself went from strength to strength, the finances of the new company were decidedly shaky and its survival was often in doubt. Looking back on this period 40 years later, Raymond recalled that he "edited the Journal with one hand, whilst discharging official duties and earning my living with professional work on the other. This was not altogether good for the Journal. I could have done better work there if I had not been forced to do so much elsewhere. On the other hand, the Journal could not support me, except that it gave me a prestige which commanded other income for its support as well as my own."

Just how near the enterprise came to complete disaster will never be known, but it must have been close. Willard Ward, the "third man" in the company partnership, wrote that Raymond used to save as much as he could from his salary as U.S. Commissioner of Mining Statistics to help out the Journal. "He would sit up late at night," said Ward, "writing stories for the *Christian Union*, stories for children and adults, even poems, and the cash he derived from all these sources would go out in a cheque on Saturday noon to old Brady, the typesetter, who had to have his money before he would let the formes out of his possession to go to the press. The rest of the office employees were, I think, regularly paid, but Dr. Raymond, whenever he got anything for his work and his advances, got stock. I don't remember to have heard of any shortness of stock!

"As for dividends, I don't know that we ever thought of any such sordid consideration. The only apology we had for existence was the enlightenment of the profession, and we tried to make everybody, directly or indirectly, interested in any mining or metallurgical enterprise believe that it was his duty to subscribe for the Journal in order that our usefulness might be continued for the benefit of future generations."

And so it was. By the mid-Seventies, the financial crisis had passed and the efforts of the partners began to bear fruit. "We are", confided an editorial, "now beginning to reap the results of a policy, so perseveringly maintained, and while we are by no means exhausted of that enthusiasm that has made sacrifice easy hitherto, we may perhaps be pardoned for expressing our particular pleasure in the present and prospective financial prosperity of the Journal."

Rothwell, with a practical approach to all problems, undoubtedly played a leading part in this revival of fortune. He also emerged from the crisis as owner of most of the stock. In 1875, he became joint editor with Raymond and thereafter devoted his time chiefly to the paper's interests. This is just what the Journal needed—someone of singleness of purpose, always available to plan and co-ordinate editorial material and steer each issue to the printers.

It is to Rothwell, above all others, that *E/MJ* owes its development in the last quarter of the 19th century. Immediately he took control he made the company's first take-over, acquiring the *Coal and Iron Record*, together with its editor, Benjamin R. Weston, and added four more pages to the Journal, devoting the extra space to coal and iron news, written by the new member of his staff. He increased the market reports, and raised prestige by publishing a series of 60 lectures on mining by Professor Sir Warrington W. Smyth, the eminent professor of the Royal School of Mines in London, who was a leading figure in its early history and a world-wide authority. In 1877, the Journal opened an office in Denver, just at the time when Colorado was coming into the news with the discoveries at Leadville and elsewhere. The paper thickened its pages again when the *Polytechnic Review* was absorbed and a new regular feature added on the progress of science and art.

With justifiable pride, the editors were able to claim that their paper had become "a mirror in which the mining industry may see a reflection of its development". It was, in fact, much more than a mere mirror; already, in the first 20 years of its life, it was playing a major part in shaping the industry which it had set out to serve. It had given a big impetus to the growth of professionalism in a pursuit which had begun as a race for wealth in which everyone could join, provided they had some primitive tools and an adventurous spirit.

Rothwell and Raymond demonstrated that the industry could prosper in the long term only by the application of the highest technology and skills. They urged the development of technical education and the establishment of a national school of mines. Just as Henry English had waged war on fraudulent companies and crooked promoters in Britain, so they took up the campaign in the United States. They were quick to condemn any action which savoured of sharp practice and they sponsored the American Bureau of Mines as a volunteer body, composed of mining engineers of high reputation, to scrutinize mine promotions and warn the public against swindlers.

Together, too, Rothwell and Raymond were mainly responsible in 1871 for launching the American Institute of Mining Engineers, and for several years *E/MJ* was its official organ.

However, as time went on it became increasingly apparent that joint editorship between the two forceful men, two men of remarkable individualism and character was becoming difficult. In such circumstances there were bound to be differences on editorial policies; the paper demanded control by one man with complete authority and responsibility. And that man had to be Rothwell. An experiment was tried by which each editor was identified with what he wrote in each issue; but it did not work and Raymond resigned his position. But this did not stop him writing for the Journal. From now on he was a "special contributor", with all his articles appearing under his own name. He continued to act in this capacity until his death 30 years later, though towards the end he accepted a general change of policy which limited by-lines to members of staff.

The success of Rothwell's earlier review of the coal industry, which had been produced in book form under the title *Statistics of Coal* encouraged the company to enter a wider field of publishing with a series of scientific and technical books. Rothwell aimed to conduct this venture not so much as a money-making department—though he obviously expected it to pay its way—but as a means to provide the profession with a succession of practical treatises on all subjects about which comprehensive and authoritative information might be required. In 1887, the Journal began publication of Howe's great work on the *Metallurgy of Steel*, which was published as a book three years later. Other volumes followed, but Rothwell became increasingly concerned with metal statistics. He was without doubt the first to recognize that the vast stores of statistical information which accumulated during the day-to-day activities of an editorial office—whether or not immediately published—could be of great importance for the intelligent direction of the industry.

By 1889, as a special Government agent for the census of that year, Rothwell had organized the collection of statistics of gold and silver in the United States, and, for several years previously, he had arranged for figures for copper, lead and some other metals to be put together and published at the end of each year in the Journal. But Rothwell was now planning something far more ambitious. The compilation of statistics of an industry rapidly expanding over the face of the globe, was a work so vast that it had hitherto been considered impossible, except through the resources of governments, and, as Rothwell put it, "because the machinery of government is not adapted to the rapid attainment of results, the statistics of the mineral industry have been so tardily collected and published in all countries that their value has been greatly impaired".

Moreover, it was not only the *machinery* of government that was responsible for such inadequate information, it was also in many cases a distinct reluctance to provide money for such a purpose. Even in the United States, the Senate in 1891 opposed the passage of an appropriations Bill allocating $50,000 for the collection of mineral statistics, saying that as a good deal of money had been spent on such an exercise in the previous year, they saw no reason to vote more than $10,000 to updating them, and they considered that major collections need be made only once every 10 years. "The absurdity of such an idea", declared an exasperated Rothwell, "should be perfectly evident to men as intelligent as these, but it must be said that they do not understand the magnitude or the importance of such work."

In the following year, Rothwell demonstrated the importance of the subject by producing a statistical supplement of 78 pages and its success led to a decision to bring out an annual publication covering the whole field of the world's production of minerals, metals, and, so far as possible, the principal derivative products, together with a review of mining and metallurgical progress from both the technical and commercial standpoints.

And so, in 1893 there appeared the first of the annual volumes which were to become famous all over the world, *The Mineral Industry—Its Statistics, Technology and Trade*. The first volume of more than 600 pages contained "statistics from the earliest times to the end of 1892", and the information was meticulously updated every year—for the next 50 years.

In a preface to the first issue, Rothwell could not refrain from underlining government reluctance to take a more positive role in such an enterprise. "Unaided by any government powers to enforce the making of returns, we have", he wrote, "relied for success solely upon personal courtesy and confidence, and upon the intelligent appreciation of the value of the work to the industry at large, and this great volume is the monument we have erected to the courtesy of those whose prompt and willing co-operation has alone rendered its success possible. Long experience of this kind of work has fully demonstrated the fact that men are in general more willing to give important and correct information to the private individual, who can be held responsible for its proper use, than to a more or less impersonal government. It is indeed extremely rare that any producer neglects or refuses to give full, truthful, and satisfactory replies to our request for information."

The names of the contributors to the first edition prove his point; the list is no less than an encyclopaedia of the great personalities of the mining world at this time—men such as James Douglas, "father of the American copper industry" and already president of the Copper Queen Company; William H. Adams, foremost authority on the pyrites and phosphates industries; William Benedict, expert in tin mining and widely experienced in Central and South America as well as in the States; John A. Church, at one time a member of *E/MJ*'s staff and who was responsible for opening up and operating many mines in the northern provinces of China; J.S. Jeans, of the British Iron and Steel Institute; James F. Kemp, professor of geology at Columbia College; Emile Delecroix, authority on the French coal industry; the industrial chemists, F.E. Engelhardt and Francis Wyatt; J. Langeloth, president of the American Metal Company; George Kunz, gem expert to Tiffany's in New York; H.O. Hofman, specialist in lead metallurgy; George A. Sonnemann, pioneer in the application of electricity to the mining industry; and many others.

Rothwell himself devoted much time to *The Mineral Industry*; the statistical side interested him the most and nothing gave him more satisfaction than to collect and publish commercially-valuable statistics of some industry for which previously there had been no figures. Few men could analyse a compilation of figures more quickly or more thoroughly, and few could reproduce them in so compact and clear a form. The annual volumes, pioneer of the statistical books which line the shelves of mining economists and engineers today, and are one of the accepted tools of the planners, represent Rothwell's greatest work during his long and distinguished editorship of *E/MJ*.

France honoured him in 1898 with the gold medal of the Société d'Encouragement pour l'Industrie National de France, given to the author—French or foreign—of the work judged to have exercised the greatest influence on French industry during the preceding six years. Credit for the business management of the project went to Mrs. Sophia Brauenlich, who became the first American woman to be elected a Fellow of the Imperial Institute of London.

Richard Rothwell died in 1902 but, for nearly half a century more, the volumes which he started continued to constitute the best encyclopaedia of current mining and metallurgy that were available.

CHAPTER SIX

Britain and Abroad in the Eighties

*B.T.A. Bell, pioneer of Canadian
mining journalism*

Henry English, founder and first editor of the *Mining Journal*, died in 1855. A modest 12-line obituary in the Journal acknowledged his "energetic interference in all matters affecting the mining interest . . . with a generous heart and active mind, he was always prominent in any endeavour to promote beneficial objects".

Although he did not reach the age of 50, English lived long enough to force government recognition of the necessity for proper safety regulations in the British mines, even though precious little progress had been made in improving the conditions in the 10 years since he first aroused the sympathy and support of the Prince Consort. He also saw the inauguration of the Royal School of Mines, for which he had campaigned so vigorously; and, of course, he had firmly established the Journal as an influential force in all affairs of the mining industry in Britain.

Some years before his death, English had taken a partner into the business, Richard Thomas Middleton, who succeeded him both as editor and general manager. Very little is known of Middleton's birth and background, except that he "enjoyed a strong constitution and robust health, which he was careful to preserve". He also appears to have been something of a diplomat, displaying, according to one report, tact in both his business and private life, during which "his timely mediation and practical suggestions in matters entirely unconnected with the Journal not only prevented litigation, but restored disputants—of whom, in connection with share transactions and the promotion of companies, there are, unfortunately, too many—to their previous position of friends".

Middleton managed to preserve his robust health—and many of his friends—until his 80th year. His death in 1883 was followed by a short period in which there is no record as to who was responsible for the paper's editorial direction. However, in 1888, George A. Ferguson assumed the editorial chair. Glancing through the file copies of his charge, he would have seen that in its first half-century it had grown from a modest eight pages to a substantial 28. He would also have noticed that its general appearance had hardly changed in that time; the format was exactly the same, though there had obviously been some face-lift in typography. Whether this continuity should be regarded as proof of the sound basis on which the publication was built in the 1830s, or a conservatism and lack of imagination in the development of its physical presentation over the years, is difficult to say. However, there was a more significant change in its editorial outlook. Whilst it was still thundering out a policy "to cleanse honest enterprise from the limpets and barnacles that cling to, and hamper the progress of, legitimate business", it had broadened its horizons and had already assumed something of an international character, both in its editorial content and, to a lesser, though ever-increasing degree, in its readership.

The reasons for this new outlook are not difficult to understand. First, the discovery of gold in California in 1848 and in Australia three years later, gave an enormous impetus to prospecting overseas and, within a few years, more than a million emigrants left Britain to seek their fortunes abroad. Later in the century the pattern was repeated in the direction of Canada, where the advent of the railway and the drive to open up the western provinces combined to create opportunities for prospectors and mining entrepreneurs; and to South Africa, where the principal magnets were the discovery of diamonds in Kimberley in 1866, and gold in the Witwatersrand 20 years later.

Those who answered such challenges became the pioneers of the new lands. Mostly they travelled alone, on foot and with a few pack animals. The mining centres of the world, active and dead, are a monument to these men, to their courage in overcoming obstacles and hardship, and to their skill in reading the signs which led them to their goal. Much has been written of the luck which sometimes led them to great discoveries, but seldom to much personal reward—a chase after a straying horse, a precious stone picked up to hurl at a marauding beast, the green stain of copper on the rock whereon lay the horns of an antelope shot for supper. But what is too often forgotten is the long and painstaking search in hostile country which ended in such discoveries—and the fact that for every bonanza there were a thousand disappointments.

A good proportion of those who set off from Britain for the promised lands in the second part of the 19th century were Cornish copper and tin miners. Mining in Cornwall had reached its peak by the late Sixties, and then the bottom fell out of the market. The dramatic change in Cornwall's fortunes has been described by many writers, but none more graphically than Daphne du Maurier in her book, *Vanishing Cornwall*: "The great days were over. Companies folded up. Mines closed down. Hundreds, then thousands, of miners found themselves out of work, without hope of employment . . . there was no alternative to starvation for the miners but mass emigration. A third of the mining population left Cornwall before the end of the century, taking their skill to other continents. Back at home, their towns and villages were left unpeopled, the mines themselves deserted, no smoke coming from the tall chimney-stacks, no sound of engines in the pumping houses, the land about them reverting once more to barren waste and scrub."

And as the "Cousin Jacks" and other pioneers roamed the world, the *Mining Journal* enlarged its coverage to describe the birth and development of the mining enterprises which were established as a result of their discoveries. The Cornishmen were first-class miners, recognized everywhere for their skill as engineers and as tireless workers. They were not gentlefolk, but to refer to them, as did the *American Journal of Mining* under its first editor, as "the most obstinate and troublesome of men to deal with" (Chapter 5) is a remark which cannot be allowed to go unchallenged. Daphne du Maurier has an explanation for their occasional misconduct: "They were a race, not of violent men, but of individualists, making their own terms when they streamed or dug for tin, beset by no ring of employers as in other industries, and this was mistrusted by those who had not the courage to do likewise. Even the tinners' festivals, customs and pastimes were either feared or frowned upon as being un-Christian—hurling, wrestling, cock-fighting—leading to excess, drunkenness and bloody fights ... The average life-span of a miner working underground was 47 years, if he survived the all-too-frequent accidents; and when he came to the surface to end his days in an overcrowded cottage, spitting black dust, existing on a diet of potatoes and barley gruel, there was no trade union to look after him, no governmental agency or private benefactor to espouse his cause."

As well as extending its coverage overseas to all territories in which mining development could be reported and commented upon, the Journal's broadening outlook was reflected in 1890 in its promotion of an international exhibition of mining and metallurgy at the Crystal Palace in London. The Palace, designed by Joseph Paxton and erected in Hyde Park for the Great Exhibition of 1851, had been dismantled and rebuilt at Sydenham, in South London. The concept of holding an exposition of world mining under its glass roof was much in keeping with the late Prince Albert's ambition to make London the centre of the scientific world and to foster goodwill among the nations. George Ferguson, now editor of the Journal, who initiated the plan, was also inspired by the success of the 1889 Paris Exhibition, and doubtless impressed by Alexander Eiffel's magnificent tower, which embodied his life-long experience in bridge-building and which still stands today, not only as the city's most famous landmark, but as a monument to 19th century engineering skill.

The time was certainly opportune to put world mining on show for the first time. The previous 20 years had seen some immense advances in technology, particularly in the metalliferous field, where extraordinary progress had been made in the economic treatment of refractory ores for the recovery of precious metals. Mining had suddenly assumed a greater importance in the world than ever before—whether as an industry, a profession or as a vehicle for speculation and investment. Ferguson and his colleagues at the *Mining Journal* found much enthusiasm for their idea to mark this moment of history. A strong executive committee was formed under the chairmanship of W. Pritchard-Morgan; Lord Fife agreed to become president; the Lord Mayor of London, Sir Henry Isaacs, became patron; and George Ferguson was appointed secretary. The list of vice-presidents read like a *Who's Who* of the top echelon of the mining industry, both professionally and commercially; it even included Sir Warrington Smyth, first professor of mining at the Royal School of Mines when it opened its doors nearly 40 years earlier, and now in ripe old retirement in the westcountry—but, alas, he was to die before the arrangements were completed.

The mining states of Australia gave enthusiastic and practical support, together with the governments of New Zealand, Mexico, Natal, Serbia, Uraguay and Honduras, whilst many of the big mining companies of Europe, the United States and South America contributed exhibits.

The exhibition was opened by Lord Thurlow before a vast crowd in the last week of July. The Colonial Secretary, Lord Knutsford, paid an early visit and noted with satisfaction the pre-eminence of exhibits from the Empire, particularly those from New South Wales, which covered an area of 15,000 square feet, occupying, as *The Times* correspondent noted, "nearly the whole of the south transept of the Palace, from the Handel Orchestra to the crystal fountain". The success of the Broken Hill enterprise in Australia was reflected in a handsome silvered column reaching to the roof and surmounted by a figure of Atlas with the world on his shoulders. Hardly original, even for those days, but highly impressive, as pictures in the Journal prove.

The big show went on for more than eight weeks and would have been extended but for the fact that the Palace was needed for a dog show. Of its success, there was not a shadow of doubt, its main effect being to direct the attention of British and foreign capital to mineral resources in all parts of the world, especially those over which the Union Jack was then fluttering proudly. Ironically, the only big disappointment was the rather meagre support given by the host country. It was, in the event, much more a "Greater Britain" exhibition than a British venture. As the *Pall Mall Gazette* remarked: "Considering that our mineral products exceed in value £73 million a year, we cannot be said to be guilty of ostentation."

Unlike most events of its kind, the exhibition even made a financial profit—more than enough to offer an honorarium to its "indefatigable secretary" which, of course, he graciously declined. His reward was the satisfaction of a job well done in the wider interests of the industry which his paper served—and the enormous volume of editorial material on which he could draw for many issues to come.

*

One afternoon in 1871 a young shorthand-writer came out of Exeter Hall in London's Strand, where he had been reporting a religious meeting, and strolled down to Fleet Street, where he was to catch a horse-bus from the corner of Chancery Lane to his home in North London.

He arrived at the corner just in time to see the bus disappear with a clatter of hooves into the distance. With a quarter of an hour to kill, he popped into a Fleet Street barber's shop to get his hair cut, emerging just in time to mount the knife-board of the next north-bound bus. As he clambered aboard, a hand touched his shoulder and a voice in his ear said: "I have seen you reporting religious meetings; can you come down to the City and report a company meeting for the *Mining World*?" The young man agreed, and so began an association which led him to become editor of the paper for more than 60 years and its sole proprietor for over 40.

William Chisholm was born of Highland stock in Glasgow in 1846, eight years after the coronation of Queen Victoria, and died during the early months of the Second World War, in his 94th year. Business brought his parents to live in London whilst he was still very young, and one of his childhood recollections was being held up in his father's arms to catch a glimpse of the funeral procession of the Duke of Wellington in 1852 as it made its way slowly up Ludgate Hill to St. Paul's.

The family being far from prosperous, William had to find himself a job as soon as he had finished school. He worked for the old West Midland Railway, first at Bridgnorth and later at Dudley. It was at

Dudley that he first became interested in shorthand; he devoted all his spare time to its study and when he was offered a job on the staff of the National Temperance League in London, his proficient note-taking proved a great advantage. Here, he was called on to provide verbatim reports of public speeches, among them by some of the great preachers of the day—Charles Spurgeon, the Baptist, whose 50 volumes of published sermons include several transcripts of Chisholm's notes; D.L. Moody, the American evangelist who toured Britain with Ira Sankey, the gospel hymn-writer; and John B. Gough. The young shorthand-writer was much influenced by Gough, who was a reformed drunkard, and in his later years Chisholm always attributed his long life and excellent health to the fact that he was a total abstainer and non-smoker.

The invitation to cover a meeting for the *Mining World*, which had begun publication only a month or two earlier, was followed by several other similar assignments and, before the year was out, Chisholm found himself on the staff. The paper had been formed by two men who had interests in the city: one was William Dunbar Quinlan; the other was a man named Biggar, of whom nothing more appears to be known. According to an editorial in the first issue on April 8, 1871, the paper was established as "an independent and thoroughly impartial organ of the various interests—mining, trading and commercial —forming the mercantile community. In its columns will be found thoughtful and well-digested articles by competent writers on all important questions affecting these interests. Public meetings will also be so fully and impartially reported as to render the *Mining World* peculiarly a shareholders' representative. In order still further to increase its value to bona fide investors, a special article will appear on the new companies registered from time to time. The articles of association will be carefully analysed and all objectionable features fearlessly exposed, and no pains will be spared on the part of the conductors to warn the unwary from embarking on disreputable schemes similar to those which in times past have brought so much discredit to joint stock enterprise."

Finally, there was an invitation to readers to contribute to the paper's correspondence columns, provided that their letters were "written in a fair and manly spirit, containing legitimate criticism".

Much of what was spelled out in that original prospectus remained good for the life of the paper, which lasted for very nearly 100 years until it finally disappeared in a torrent of take-overs and mergers in the late 1960s. *Mining World* always remained very much more a paper for mining investors than for mining engineers, few of whom ever even saw it, let alone read it with any regularity. Largely because of this, *Mining Journal* did not regard the new venture as a serious competitor despite the fact that basically the subjects the two papers dealt with were very much the same, even though they approached them from different angles. It would be naïve to suggest therefore that *Mining Journal* did no more than "take note of" the presence on its doorstep of a paper with such stated aims and objects; indeed, it is possible to detect in the Journal about this time an increasing interest in the financial aspects of mining and rather less emphasis than previously shown on matters of direct concern to the professional mining man and metallurgist.

Certainly, in the 1870s there was great demand for verbatim reports of conferences, gatherings of learned societies and joint-stock companies. Whereas *Mining Journal* had always given considerable space to official company reports—and continues to do so to this day, both editorially and as paid advertising—*Mining World* concentrated more on the impromptu speeches and observations made by directors and shareholders at company meetings. This was something new at this time and consequently the paper became popular among those who were looking for a more critical analysis of mining company activities—and, let it also be said, with those who attended such meetings and liked to see their names and comments in print. And *Mining World*, now with one of the best shorthand-writers in the country on its staff, was able to provide such a service.

Horse-buses seem to have played a significant part in *Mining World's* evolution for it was on the top deck of such a vehicle that the name of the paper was decided. Apparently, the partners had been considerably exercised with the problem of giving their venture an appropriate title and had reached no conclusion until one day when

71

they were swaying in the bus down Fleet Street to their office, they saw an advertisement for the *Christian World*. After that there was no argument: what was good enough for Christianity was good enough for financial mining.

Little is known of the first editor, whose name was Brodie; nor of his successor, named Heron; but neither of them remained for more than a few months, and William Chisholm quickly moved into the chair that he was to occupy for the next 60 years. In 1888, Quinlan took him into partnership and when the former died six years later Chisholm became sole proprietor, continuing ownership until his retirement in 1932, when he handed over the business to his two sons.

When *Mining World* was launched, Europe was in the shadow of the Franco-German War, but happily its effects were not seriously felt in Britain and when Chisholm became editor he set to work to shape the paper's character in strict conformity with its prospectus. He always expounded the truism that the finest mine in the world was worthless if it lacked capital to develop it properly. The famous mines of the Kolar field, in Mysore State in India; the "Golden Mile" in Western Australia; and the Ashanti goldfields on the West Coast of Africa, were all favourites of his and the recommendations which he made to his readers to support these companies turned out to be good advice.

Mining World always liked to feel that it was competitive with the contemporary Press and claimed something of a scoop in giving advance warning of the notorious Jameson Raid in South Africa in 1895. An article in its issue of December 21 of that year, under the heading "Is it revolution?" declared: "If Mr. Rhodes has 1,000 such men as won the Matabele War on the northern borders of Transvaal, as he is said to have, and intends to use them unless justice is rendered to the foreign population of Transvaal next month, and if the great financial houses really decide to throw their influence and their wealth into the scale against President Kruger, there will undoubtedly be a revolution, and that before another six weeks passes over our heads."

Two issues later the *Mining World* reported the Raid and its disastrous consequences, but with a modicum of sympathy for "the gallant doctor (Jameson) who dared so much for his country and who

will have to pay the penalty because he has not been successful ...
deserted by his friends, cast over by the Chartered Company, and
repudiated by the Home Government, the result was inevitable."

It is interesting to note that in direct contrast, *Mining Journal* could
find no sympathy in its heart for the unfortunate doctor. "Could any
filibustering be more audacious, more insane?", its editorial asked.
"He has forfeited all sympathy ... not only has he dragged the British
flag through the dirt, not only has he given a blow to British prestige,
but he has brought down upon us the indignation and anger of the
world."

Perhaps in these opposing attitudes to the Jameson Raid lay the
fundamental difference between the two papers. It may be difficult to
claim for *Mining World* that it played any significant part in the
development of the mining industry, but it certainly contributed to a
better understanding of the way in which mining companies ran their
business, and it kept a watchful eye for those which dazzled only to
deceive. It must be remembered, too, that it was born at a time when
there were few papers devoting much space to investment in mining,
or in fact in industry generally, and it is therefore deserving of a place
in the records as a pioneer of modern financial journalism.

*

Although silver and copper were discovered during Champlain's
explorations around the St. Lawrence River nearly 400 years ago, it
was not until the second half of the 19th century that mining became
of economic significance in Canada. As in California, the consolidation
of sporadic discoveries into an industry of national importance was
preceded by some glamorous years, the history of which is marked by
such place-names as Barkerville, in the Cariboo district of British
Columbia, scene of a wild gold rush in mid-century; the Fraser River
gold finds; the Kootenay region, in which the rich silver-lead-zinc
Sullivan mine was found; Sudbury, where blasting operations for the
Canadian Pacific Railway revealed the great nickel-copper deposits
of the Sudbury Basin; and Klondike, in far-off Yukon, which set off
one of the great gold rushes of the century.

These were but a few of the early highlights for, as the century grew old, mineral discovery and development gathered momentum in an ever-widening spectrum across the country—gold and coal in Nova Scotia; coal in British Columbia; copper and lead in Newfoundland; and large asbestos finds in Quebec's Eastern Townships.

As the 20th century dawned, a new mining frontier began to be opened up on the Pre-Cambrian shield of northern Ontario and Quebec. The Cobalt strike of 1903 touched off a rush of prospectors in search of the precious and base metals of the shield, where a whole new complex of primary and secondary industry sprang into being, even though Canadian prosperity continued to depend in the main on its western wheat fields.

Mining journalism in Canada is a product of these times, even though its origins were parochial and its early objects somewhat personal. In 1879, the *Canadian Mining Review* (renamed the *Journal* from 1908) was founded by W.A. Allan, a successful entrepreneur in the phosphate mines in the Ottawa area, and Grant Powell, a civil servant and future Under-Secretary of State.

C.M. Percy, the English writer of *The Science and Art of Mining*, described the *Review* in the mid-Eighties as "a diminutive pink sheet of eight pages, principally devoted to the booming properties of its then owner". Whilst Mr. Allan and his partner were obviously aware of the fact that the mineral wealth of the country was now being opened up from the Atlantic to the Pacific, they continued to direct the interest of their readers mainly to the Ottawa phosphates business—and to the activities of Mr. Allan. In June, 1883, for example, the *Review* congratulated Mr. Allan on acquiring the Emerald Mine, "the most valuable phosphate mine yet discovered in Canada". Within a couple of months it again showered him with congratulations, this time for selling the same mine "to a company of American gentlemen for the splendid sum of $125,000". A few weeks later it drew attention to his "unerring judgment, his energy and the business-like manner in which he has always conducted his operations", and in the following March it reported new purchases by Mr. Allan, "displaying his usual good judgment and foresight".

Obviously, this continual self-eulogy became a bit of a strain on

Mr. Allan and when his partner, Mr. Powell, was called to higher office, he began to look around for someone who would take charge of the paper editorially and give him more time for his dealings in phosphates. He found his man playing in a cricket match in Ottawa in 1886, and engaged him hardly before he could get out of his flannels. His name was B.T.A. Bell and his appointment as editor of the *Review* was the best thing that Allan did for Canadian mining journalism.

Whatever qualifications Bell may, or may not, have had for such an appointment, there was no doubt about his prowess as a cricketer. Playing for the Canadian national team, he had made the highest aggregate score in the country that season—1,306 runs in 36 innings—and had been chosen to tour England. His journalistic experience had apparently been limited to a brief editorship of the (Canadian) *Athletic News* and some freelance writing for the Montreal newspapers.

But, at 26, Bell was already a man of the world with a rich experience of life. Born in Edinburgh, Scotland, he was educated at Stewart's College where, according to one report, he "developed that independence of thought and quickness of action which characterized him as a man and was early shown by his decision that *seeing life* was better than *schooling*". As soon as he left school he was off to farm in Canada. Landing at Halifax, he made his way with four friends to Brandon, where they lived in a tent on the prairie outside the town until their "shack" was ready. But Bell did not take to farming. He got a job in the freight department of the Canadian Pacific Railway at Winnipeg. One day, on a visit to Rat Portage, the Manitoba contingent of the Canadian Voyageurs marched through the town on their way to attempt the relief of General Gordon at Khartoum and, seeing some of his old chums in the ranks, Bell promptly joined them. He received the Khedive Star for service in the Sudan, and, on returning to Canada, he was attached as a lieutenant to the Prince of Wales Rifles. The Rifles were then on stand-by for the Riel Rebellion, but they were never ordered into action and Bell began to make a career for himself in sporting journalism.

It obviously took little persuasion to get him to switch his interest to mining and, from the start of his editorship of the *Review*, he showed a profound concern for the welfare of the industry. Like Henry English in London, 50 years before him, he at once presented himself as a crusader against corruption and dishonest promotion. Bell's targets were the creatures of the Canadian frontier, lured by a booming industry and living off the unsuspecting investor. Among them was Charles Lionais, who sent Bell a Press notice about some new asbestos properties in the Eastern Townships. Bell published the announcement, but warned: "Perhaps it might be well to ascertain and prove the true value of the properties before going to all this expense. Extravagant and ill-advised management has in the past wrought enough harm to our Canadian mines as a number of properties with which Mr. Lionais is not unacquainted can testify." Lionais complained that "owing to the extensive circulation of the *Review* in Great Britain and the United States, it will have the effect of deterring capitalists from investing their money in this property", and threatened a lawsuit if the offending remarks were not withdrawn. "We will do nothing of the sort", said Bell, "and if Mr. Lionais is really serious and is not catering for a little cheap notoriety, we will meet him in court and substantiate any opinion we may have expressed in these columns." Nothing further came of the case.

Bell's "Dr. Moriarty" was Charles Miles de Tracy Dobson, "a fraud, deadbeat and imposter of the very worst character". Dobson posed as an associate of the Royal School of Mines and bore credentials from an eminent firm of London engineers. On his arrival in Ottawa, Dobson placed professional advertisements in the *Review* before Bell caught on. Before he was exposed he had left a trail of swindled investors. On another occasion the Sydney and Louisbourg Coal and Railway Company used the excuse of ice damage to its piers in order to justify a large financial loss to its shareholders. The ever-suspicious Bell produced a letter from the Cape Breton Board of Trade denying any "impact of ice" and asserting that "such a report is a gross libel on the fair name of our magnificent harbour".

Bell's philosophy about speculation and investment was clear-cut: "Profits coming from the legitimate working of mines benefit the industry by inducing further investments of capital, but the speculator who sells a prospect for 10 times or 100 times its value is no benefactor.

It is only by making mining profitable to those whose money is invested in the actual working of the mines that a healthy and legitimate growth of the mining industry can be attained."

By this time Bell had no longer to look over his shoulder for Mr. Allan for, within a year or two, he had become proprietor as well as editor. Mr. Percy, of *The Science and Art of Mining*, who had been critical of "the diminutive pink sheet" a year or two earlier, was now applauding Bell for "taking hold of the literary reins, entirely changing the policy and working the paper as a purely journalistic enterprise". He added: "During the time that Mr. Bell has had control he has been honoured with about one action for libel each year, every one of which has been abandoned before going into court. An upright editor has his troubles and suffers a good deal of wicked annoyance and unrighteous expense . . . I pronounce the *Review* a mining journal not surpassed in the English language, nor any other language. The Canadian mining community are justly proud of it, and we at home in England highly appreciate it."

Editors of other contemporaries paid tribute to the *Review's* fearlessness. The *Australian Mining Standard* noted that the paper was "at one with ourselves in the exposure of mining swindling and swindlers" and added, a little sadly: "unfortunately for the public who put their trust in these colonies, the action of the Press, which often has the information necessary to enable it to issue specific warnings, is trammelled by absurd and unjust laws, and we admire outspokenness such as that now referred to, but we are not able to follow such a courageous example without incurring ruinous penalties."

Bell's exceptional concern with those whose activities could damage the young industry quickly earned him an impressive reputation. He was recognized as a spokesman for mining who was above the influence of regional, sectional or other "special interests". He emerged as the central figure in the industry's first major battles against discriminatory taxation and government interference. Canada's first professional mining body, the Gold Miners' Association of Nova Scotia, adopted the *Review* as its official organ and, in 1889, Bell initiated an invitation to the American Institute of Mining Engineers to hold its autumn meeting in Ottawa. The invitation was warmly accepted: there was a welcoming address by the Prime Minister, Sir John A. MacDonald, and the programme included the delivery of technical papers, and field visits to a number of mining areas, including Sudbury, the Eastern Townships, and to the Little Rapids phosphates mine—which, curiously enough, had been acquired by the redoubtable Mr. Allan.

But life for Bell was not all crusading and conferences. His colourful attacks on the shadier side of the industry make fascinating reading, but in the main the *Review* focussed on the legitimate mining operations in a period which saw the emergence of the Sudbury mining camp, the discovery of mines in the Trail-Rossland area of British Columbia, and the first placer-gold in the Klondike. In five years he transformed the paper from eight editorial pages and four of advertising, to 50 pages, 29 of which were reading matter, and increased circulation from 400 copies to 4,000.

Reports on the Sudbury operation appeared in practically every issue and after one attempt at calculating the value of the great basin, Bell concluded: "These figures are so large that we can scarcely realize them, but they will at all events serve to demonstrate the value of the industry which was started only four years ago at Sudbury. Recent experiments with alloys of iron and steel have made clear the fact that Ontario is wondrously rich in her mineral possessions, and if a proper spirit of enterprise is exhibited, she may look forward with confidence to occupying a foremost place among the countries of the earth in the production of economic minerals. She has great supplies of rich iron ores as well as ores of nickel. It will be the fault of her own people if she does not become a seat of the world's nickel-steel industry."

Prophecy was not the least of Bell's accomplishments, and he remained not without honour—even in his own country.

CHAPTER SEVEN

Fathering the Institutions

*George Seymour, jnr., First
President of the IMM*

The quickening pace of scientific and industrial progress in the more developed parts of the world in the second part of the 19th century produced an increasing number of highly skilled men and women of many disciplines. A new "professional class" emerged from a society which—in Britain, at least—had been broadly distinguished by landowners and capitalists at one end of the scale, and "workers" at the other.

Predominant among the new professionals were engineers, chemists, physicists, architects, designers, medical men and teachers who, in the course of time, tended to organize themselves into institutions and associations for the discussion and promotion of their different interests.

Those pursuing careers in mining and metallurgy were no exception: between 1870 and the turn of the century, professional bodies concerned with their special affairs were set up in at least four countries—the United States, Britain, Canada and South Africa; in three of these countries, the chief stimulus, if not the original idea, came from the mining Press, and in each case early administration was provided by mining journalists.

The precise reasons for the formation of these institutions are not always clear from the printed records, or from the original articles of association, and certainly the declared motives of the founders differed considerably from country to country. In the United States, where the inauguration of an institute followed the development of vein mining, which demanded competent engineers instead of casual entrepreneurs, the overriding purpose was to secure a wider dissemination of professional knowledge. In England, the principal concern appears to have been to establish professional standards for the British engineer working overseas, and particularly in the Empire, and to set him apart from the "rag, tag and bobtail", by which was probably meant those who had not been to the Royal School of Mines or Camborne, and others who had the temerity to pose as "engineers" without recognized qualifications. In Canada, the object seems to have been different again: here, the original idea was to merge the various regional associations into a sort of "national protection society" for those engaged in the industry—a body to combat discriminatory taxation and other government policies which might be a threat to the mining business. In other words, the Canadian institute appears to have been constituted in the first instance to perform primarily those functions of self-preservation and government liaison which are today carried out by the Mining Association of Canada.

In the United States, the American Institute of Mining Engineers (AIME) was the brain-child of the *Engineering and Mining Journal*, and more particularly of Richard Rothwell. With two others—Eckley B. Coxe and Martin Coryell—he signed a circular letter which was sent out early in 1871 to a long list of mining men across the continent, from Pennsylvania to California, calling a meeting at Wilkes-Barre. Rothwell took the chair and Rossiter Raymond, who had been persuaded by Rothwell to go along, found himself inspanned as temporary secretary. A packed meeting of distinguished mining personalities decided there and then to launch the Institute.

The project was an immediate success. Four years later, *E/MJ* claimed its credit in an editorial: "We may say, without undue vanity and without fear of contradiction, that the extraordinary success of the AIME has been largely due to the agency of this paper. The very origin of the Institute, as its founders will confess, was a result of that common feeling which the endeavours of this Journal have aroused among the mining engineers and metallurgists of the country; and while we gladly acknowledge an indebtedness to that body for the important and interesting papers with which the members have enriched our pages, we do not hesitate to claim its great success as part of the fruit of our own earnest toil."

Rothwell and Raymond did, in fact, share most of the early organization of the Institute and Raymond soon became deeply absorbed in its affairs. He was president from 1872 to 1875, and in 1884 he became secretary, which at that time meant that he was chief administrator, and he continued to serve in this capacity for the next 28 years—until he was well into his seventies.

Raymond often reflected that the formation of AIME was the beginning of the end of his relationship with Rothwell in the editorial direction of *E/MJ*, though he always emphasized that there was never any end to their personal friendship. It was, however, curious that Rothwell, who was the prime mover in the founding of the Institute, devoted himself in later years almost entirely to the Journal, whereas Raymond, who shaped the early development of the paper and fed it material of such high standard, was to give so much of his life to the Institute. Raymond never thought that the Journal suffered from this exchange and always declared that he was as heartily proud of its subsequent success as if he had more to do with it.

There never was any doubt in Raymond's mind about what the prime function of the Institute should be. His early instincts as an editor and publicist convinced him that it should play the role of a communicator of knowledge and experience and, more especially, of technological development. He therefore accepted with particular enthusiasm that part of his job which concerned the production of the Institute's "transactions", and for the next quarter of a century the immense annual volumes were essentially his work. He stimulated contributions on all subjects which he considered to be of professional benefit. He accepted what was voluntarily offered as raw material, hewed out by axe and adze. He was the joiner, carver and polisher. But he would work on nothing but good material and he made sure of its soundness before he put his hand to it at all. His erudition and extraordinary memory made him a keen and profound critic, and it was not often that anything shoddy escaped his eye.

Raymond was often called to explain the style of his editing. A story is told that the president of a mining company, observing the publication of a paper in elegant style over the signature of one of his superintendents, whom he knew to be incapable of such prose, confronted Raymond with, "Why are you trying to pull the wool over our eyes? Bill never wrote that paper; he couldn't write it if he tried." "No," replied Raymond, "Bill did not write it, but he did the work that the paper told about. It was good work and information about it is important to the industry. It is my job to present such work in a way that it can be read easily and understandingly."

This story epitomizes the philosophy of Raymond's editing. In pursuing it, he added more to the archives of mining and metallurgy than any man before him—and probably any man since—not so much by his own contributions, though they were important, but as the extractor and the producer of records and views of practical men who had not his gift of expression.

*

If it is accepted that the foundation of the American Institute of Mining Engineers was a product of the more extensive vein-mining in the United States, it would be equally true to say that the Institution of Mining and Metallurgy in London owes its origin to the development of gold mining in India, which demonstrated, perhaps for the first time in the United Kingdom, the importance of engineers and metallurgists of high professional standard in the successful exploitation of the mineral resources of the British Empire.

The opening up and development of the famous Kolar gold field in Mysore was in the hands of the highly professional engineering firm of John Taylor and Sons, and it was due chiefly to the soundness of their technical policy that, as the years passed, the mines were able to go deeper and deeper and, at the same time, by continual improvement in mining methods, it was possible to offset the consequential rise in cost by increased production.

Among those who first recognized the need to provide for the continuity of such high standards was George Seymour, jnr., a graduate of the Royal School of Mines and winner of the 1875 Murchison

Medal, and partner in the firm of Bainbridge, Seymour and Company, consulting engineers with world-wide connections with metalliferous mining. An orphan, brought up by an uncle, who was a well-known ship-owner, Seymour was trained in England and France, and his first appointment was to the Spanish Mineral Phosphate Company at Caceres. He later made several expeditions to Iceland, the first of which nearly ended in disaster when his ship was wrecked on its shore.

Seymour made an attempt to form an institution in 1886, when he arranged a meeting which was attended by some of the most eminent and influential members of the profession, including Sir Warrington Smyth, John Arthur Phillips, Professor Huntington, T.J. Bewick and Hilary Bauerman. But, for various reasons, it came to nothing. Four years later, the success of the International Exhibition of Mining and Metallurgy at the Crystal Palace (Chapter 6) again focussed attention on the need for such an institution. Indeed, it was always the intention of the Exhibition organizers that one-half of any financial surplus should be devoted either to founding a scholarship or to "helping some other institution connected with mining and metallurgy".

With this in mind—and £300 in the kitty earmarked for such a purpose—George Ferguson, editor of the *Mining Journal*, who had been organizing secretary of the Exhibition, invited a number of people to a meeting at the Journal's offices in February 1891 to discuss the subject. In recognition of his previous efforts, George Seymour was asked to occupy the chair and, once again, the big names were there, including John Taylor, C. Algernon Moreing, C.A. Alford, Arthur Claudet, Professor Huntington, Mansfeldt H. Mills (mining engineer to the Duke of Rutland), and S.H. Cox (a former member of the Royal Commission from the New South Wales Government to the Crystal Palace Exhibition).

It was made pretty clear from the outset that the intention was to confine any institution which might be formed to those engaged in metalliferous mining; there would be no truck with those who were wholly concerned with coal mining, for it was held that by the weight of their numbers they could swamp any professional organization and, in any case they already had some thriving institutions of their own in the north of England, the Midlands and South Wales. The big problem seemed to be the precise qualifications which would be demanded of members, and here Mr. Seymour found himself on rather delicate ground. "There are", he remarked, "several good honest mining engineers in London, but there are several who live, so to speak, on the fringe of the profession and who would be the first to cry down the institution and all connected with it if they were not admitted." Another speaker warned that, unless some rigid rules were adopted, anyone who had been connected with mining undertakings could join, and the institution "would thus be associated with the names of carpenters, journeymen, bricklayers, tailors and such people as they had in Johannesburg who brought disgrace upon the profession".

Other problems were fully debated and the inevitable committee—with George Ferguson as secretary—was formed to go further into the matter. The committee took a bit of time to crystallize its thinking but, held together by Ferguson and Seymour, it did its homework, and on February 13, 1892 some firm proposals were put forward at another, and much larger, meeting at Winchester Hall. Here, Seymour, who again assumed the chair, stressed the fact that London, the centre from which so much of the world's mining was financed, was the only capital without an institution for engineers in metalliferous mining. The AIME, he said, was in itself a reproach and an example to them: "Vast as are the United States, their aggregate area is relatively small when compared with the area controlled by the Ruler of the British Empire . . . they cannot compare with our imperial mining concerns. And yet we must admit that in science, in management, and in metalliferous and metallurgical science generally, they are second to none, whilst in some branches, owing, it is true, to the occurrence within their limits of ores which are practically non-existent in the United Kingdom, they have attained to a relative superiority. This condition of things it should be our desire and ambition to rectify, and it is this desire which leads me to ask you for your assistance in the embodiment of that which will, I trust, prove a permanent and mutually beneficial institution."

The outcome was a unanimous decision to form the Institution of Mining and Metallurgy (IMM) with temporary offices at the *Mining*

Journal, 18 Finch Lane, EC., and with George Ferguson as hon. secretary *pro tem.* Its object was defined briefly as "the general advancement of mining and metallurgical science, and, more particularly, for promoting the acquisition of that species of knowledge which constitutes the profession of a mining engineer". Its constitution was divided into four classes: members, honorary members, associates and students, with some well-defined guidelines laid down for the council to determine whether or not a candidate should be admitted.

George Seymour became the first president and in his first presidential address, in the lecture room of the Museum of Practical Geology in Jermyn Street, on May 18, 1892, he very appropriately spoke of the influence of mining and metallurgy on the development of civilization, and reviewed its progress throughout the world.

The IMM and mining generally owes much to both George Ferguson and George Seymour. The former has left his mark in the volumes of the paper which he edited in the closing years of the last century; the latter as first of a long line of presidents of the Institution which has now served the interests of the profession for more than 80 years.

The memory of George Seymour was recently revived in a delightful way by the publication by *Mining Journal* of some pen sketches which he made of Cornish mining characters 100 years ago. Collected under the title of *The Man Machine*, the little volume was issued on the occasion of an International Exhibition of Mining Machinery in Birmingham in 1977.

*

In 1890, a "very pernicious and unjust" measure was proposed in a speech from the throne of the Quebec Legislature. Under this Bill, all mines, "already alienated from the Crown", would be required to pay a government royalty of 3 per cent on the gross value of their products.

The Bill was the work of the French-Canadian government of Honoré Mercier, which had swept into power a year or two earlier on a tide of racial and religious feeling following the execution for treason of Louis Riel, the rebel leader of discontents in the western regions of the country.

B.T.A. Bell, the irrepressible editor of the *Canadian Mining Review*, caught the whiff of battle in his nostrils and reacted swiftly with an editorial, in which he pointed out that investors had to be *coaxed* into mining more than into any other industry for, though its profits were often large, it had a great proportion of losses. "If any tax is to be levied", he declared, "we would suggest that it should be placed upon the mineral lands held in idleness by speculators . . . For the interests of the country, therefore, we say 'Hands off' from every struggling industry, and let not the parent be the one to heap unnecessary burdens upon the child."

Asbestos producers and other mine operators had an emergency meeting in Sherbrooke, at which the proposal was condemned as "a serious menace to the mining industries of the province", and a meeting of phosphate miners was called in Montreal, where "notwithstanding that it was the closing of the shipping season and consequently a very busy time at the mines, the attendance was good". The Montreal meeting passed resolutions condemning the Quebec government, and Bell jumped to his feet to propose that "the time is now ripe to form a General Mining Association for the Province of Quebec . . . an association which would always be prepared to cope with any emergencies that might arise in the future". The resolution was carried at a re-convened meeting in January the following year, "marking", as Bell wrote in an editorial, "an epoch in the history of our mining industry." Needless to say, Bell readily undertook the duties of secretary of the new association.

On February 11, 1891, the committee of the GMAPQ secured a meeting with Premier Mercier. Bell spoke of the injurious effect of the royalty law in checking enterprise. Competition from the newly-discovered phosphate fields in Florida, he said, was already a serious threat to Canadian industry and the imposition of further burdens would discourage continued development. Mercier promised "full attention to the representations made". But the delegation was far from satisfied, and the following month Bell noted in the Review that the Premier had gone to Europe without doing anything about the Association's protests.

Meanwhile, the new body got down to other tasks. In May, a general meeting and dinner was the occasion for a discussion on mine safety, about which Bell remarked, in the tradition of Henry English, that it was a "notorious fact that in this province no record is kept of accidents, and no enquiry or report is made regarding them by the present so-called Mining Service of the Government". At the first annual Association dinner, Bell toasted "Kindred Associations", his mind perhaps already assembling a national federation. Sir William Dawson, principal of McGill University, took the opportunity to praise "the energetic secretary" and to note that the Review could be considered "a great educational power".

The campaign against "the obnoxious law" proceeded vigorously, the Association's members no doubt encouraged by the mounting scandal arising from allegations that subsidies, voted by the Government for the railways, had been applied for political purposes. Before the year was out, Mercier had been dismissed and early the following year his party was decisively defeated at the polls, being replaced by the Conservatives, led by de Boucherville.

Bell and his colleagues sought an early meeting with E.J. Flynn, the new Commissioner of Crown Lands, and Attorney General Pelletier. As a result, the Mercier law was repealed in July and the mining industry's friendship with Mr. Flynn, who later became Quebec Premier in 1896 and 1897, was cemented. Flynn was a frequent guest at the Association's functions and, years later, he was glowingly referred to as "the father of the whole mining industry in Quebec".

Inspired by the GMAPQ's achievements, the Mining Society of Nova Scotia was formed under the chairmanship of John E. Hardman and, two months later, the local Gold Miners' Association voted to amalgamate with it, passing a vote of thanks to Mr. Bell "for his efforts in that behalf, and to several Members of Parliament who have aided in this good work".

In 1893, Bell was again one of the prime movers in arranging an International Mining Convention in Montreal, an occasion at which the social niceties appeared to attract as much attention as the business sessions. The crowd was estimated at 600, including, as the Review noted, "a goodly attendance of ladies, many of them in evening dress, and the conventional dress suit was as prevalent and as acceptable as buds on a maple tree in April. The proceedings were enlivened by an excellent selection of music given by the full band of the 1st Victoria Rifles."

With a Nova Scotia association now firmly established and enjoying a good relationship with its Quebec colleagues, Bell's vision of a national body was becoming clearer. In Ontario, an attempt to form an organization had been made in 1892, but Bell would have little to do with it, maintaining that "few of its members are known to us as mining men". "There are", he wrote, "a few lawyers, three or four doctors, half a dozen speculators in mining lands, two or three land surveyors, and some others, whom the lawyers and speculators employ at odd intervals to explore the woods for minerals after fly time. Within a very narrow circle they are known as mining brokers, who buy and sell lands of others at the best commission they can bargain for, or as men who, by some chance, have got hold of a good location at a low figure and are doing their utmost to make a fortune by the sale of it."

One of the members whom Bell named, threatened him with a lawsuit, but the editor quickly disposed of him and initiated a meeting in Toronto, at which a new association was born, with James Conmee, MP for Port Arthur, as president—and Bell as secretary.

Bell was now in the extraordinary position of being secretary of all three associations—Quebec, Nova Scotia and Ontario. Federation was not far off. In July, 1894, the Quebec association chartered a steamer to take its members to Sydney, Nova Scotia, for their summer conference and, at a united meeting, a motion to federate and to invite Ontario to join was carried "without a dissenting vote". Bell celebrated this triumph by organizing a cricket match as a diversion from the usual technical tours and, according to one observer, went to the wicket "decked out in a pair of flannels which had once been white". They could, of course, have been the same pair he had worn eight years earlier when Mr. Allan spotted him in the field at Ottawa and offered him his job on the Review, for he could have had precious little time for cricket in the intervening years.

On January 10, 1896, the first meeting of the Federated Board took place in Quebec and, a year later, the new body held its first Inter-

Provincial Conference of Mining Engineers in the presence of the Governor General, Lord Aberdeen. The proceedings included Bell's report on finances and general activities, and discussions about the imposition of tariffs on mining machinery, a proposed mining museum for Montreal and amendments to the Ontario Mining Act.

Later that year, the British Columbia Association of Mining Engineers was inaugurated and agreed to federate, thus giving the movement a truly national character, and in early 1898, the Federal Parliament passed an Act incorporating the Canadian Mining Institute. At its first meeting, John Hardman was elected president and, very naturally, Bell was appointed to his favourite post as secretary.

Inevitably, with Bell steering its course, the early days of what is now the Canadian Institute of Mining and Metallurgy (CIM) were marked with some lively debates and spectacular conventions. Bell found a frequent opponent to some of his policies in a namesake, Dr. Robert Bell, of the Geological Survey of Canada, which had been founded some 50 years earlier by Sir William E. Logan, an untiring geologist, even though he was regarded by many as somewhat eccentric. The two Bells clashed over a variety of subjects, including a move—which B.T.A. Bell supported—to set up a federal Department of Mines, and the collection and publication of mining statistics.

In the summer of 1900, there was a real jamboree when the CIM and the American Institute of Mining Engineers held a joint excursion and convention in Nova Scotia, preceded by much junketing in Quebec. From Halifax, several days were spent visiting the Nova Scotia gold mines, the delegates returning "amid salvos of cheers and the greatest enthusiasm". The whole trip lasted a fortnight and, for those whose presence was not vital to the running of their own operations, there was a one-week side trip to Bell Island, Newfoundland, and the Wabana iron ore mine. The success of this, and other excursions initiated by CIM, undoubtedly inspired the organization of similar events in the ensuing years and set a pattern which has been followed broadly by international and Commonwealth mining conventions to this day.

Bell's career came to an abrupt and tragic end one day early in 1904 when, returning from lunch, he fell down the lift shaft of the

Ottawa building which housed both the Review offices and the CIM library, which he had established there. He was only 43. When he joined the Review 18 years earlier, the Canadian mining industry was in its infancy. When he died it was on the threshold of phenomenal growth, the silver discoveries at Cobalt in 1903 and 1904 providing the key which was to unlock the Canadian shield.

William Schabas, in a centennial supplement to the *Canadian Mining Journal* in 1978, wrote that Bell never discovered a mine, never worked in one, and never owned one. But it would be hard to picture the Canadian mining industry in its early days without him. And if there is any doubt about his leading part in the founding of its professional institute one has only to refer to an account, years later, in the CIM transactions, in which it is explained that though the Institute's formal beginnings were in 1898, "the germ of the idea was conceived long before then in the active brain of a man to whose strong personality, energy and public spiritness we, as an organization, owe so much".

That man was Benjamin Taylor A. Bell.

*

The sudden death of the Review's editor was followed by a period of turmoil in the ownership and management of the magazine. The editorial offices were moved from Ottawa to Montreal, and H. Mortimer-Lamb, former editor of British Columbia's *Mining Record*, was appointed to the chair; he also assumed the position of secretary of the Institute, thus carrying on the traditional ties between the Review and CIM, an association which was further strengthened when the Review was named the Institute's "official organ".

What exactly was going on behind the scenes at this time is difficult to unravel, but it would appear that American interests made a determined bid to obtain control of the Review and, in order to prevent this, a number of well-known Canadian mining men got together and put up money for a new company. Most of the finance invested was "Cobalt money" and the contributors included Alexander Longwell and Professor S.F. Kirkpatrick, who were among the discoverers of Cobalt; Dr. Milton Hersey, the metallurgist who first recognized silver in the Cobalt ore; J.B. Tyrrell, a renowned geologist and a future president of CIM; James Douglas, who was already a Phelps Dodge partner in New York and was soon to become president of the corporation; Eugene Coste, a pioneer of the development of Alberta's natural gas resources; and L.H. Timmins, of the famous family which dominated Canadian mining for much of the 20th century. Another sponsor was J.J. Harpell, who became the publication's business manager.

The new owners promptly changed the name of the paper to the *Canadian Mining Journal* and, in doing so, incurred the wrath of the *Mining Journal* in London, who accused them of "an unfriendly act" in adopting such a title, and remarked that the new paper was none other than "our old friend, the *Canadian Mining Review*, masquerading in new clothes".

Nevertheless, a new emphasis appeared to emerge under the new title. An editorial declared: "in making itself an organic part of Canada's industrial life, the Journal will strive to follow intimately the rising generation of technical men. There are few who realize the valuable work which our techincal colleges are doing in supplying, year after year, highly-trained engineers, chemists and metallurgists." This underlining of the importance of the professional man was to have a considerable effect on both the magazine and the Institute, and produced much in-fighting between the proteges of Bell, who were not technical men, and those who sought to put more technically-minded officers into the CIM hierarchy.

An unusual situation also arose in regard to the editorship. Two names appeared under the masthead of the early issues of the Journal, Mortimer-Lamb, the direct successor of Bell, and J.C. Murray. Mortimer-Lamb, it will be recalled, had also been appointed secretary of the Institute. It was therefore a little surprising when there appeared in the Journal—only a month or two after the change of ownership, and with both editors' names still under the masthead—an article highly critical of the management of CIM, and suggesting that the offices of secretary and treasurer should be amalgamated. "Without doubt", wrote editor Murray, "the Institute needs revivifying. Equally, the office of secretary is not clothed with sufficient authority or discretion ... there is great need for injecting new life into the Institute."

Editor Mortimer-Lamb promptly replied to editor Murray that he had "read with astonishment"—as well he might in his own paper—"that the *Canadian Mining Journal* which, at the special request of its publishers, was constituted this Spring the official organ of the Canadian Mining Institue, should be guilty of insinuations which are not only largely unwarrantable, but are likely to be injurious to the organization whose interests you are expected to conserve ... If I were not secretary I would go for you without gloves."

Editor Mortimer-Lamb demanded that editor Murray modify his views, but the latter did nothing of the sort and backed the candidature of H.E.T. Haultain and Milton Hersey for the positions of secretary and treasurer respectively of the Institute. Like Hersey, Haultain was a technical man, whereas both the incumbents were administrators with no professional mining background. At the annual general meeting the incumbents were returned; Mortimer-Lamb's name disappeared from *CMJ*'s masthead; the Journal lost its status as "official organ"—and CIM started its own *Bulletin.*

Although the breach was partially healed some months later and Mortimer-Lamb's name reappeared in the pages of the Journal as "special contributor", the fact that the Institute had set up its own paper continued to rankle with the Journal. The controversy reached boiling point again 15 years later, in 1923, when *CMJ* expressed concern that the *CIM Bulletin*, far from being merely a reprint of the Institute's papers, had become a serious competitor in a field which the Journal had previously regarded as its own. Not only did the Journal feel itself "wounded in the house of its friends", but it professed concern that the *Bulletin* was absorbing more and more of the secretary's time to the exclusion of other important duties, and it asserted that the Institute had nearly bankrupted itself by spending more than half its income on publications.

The Journal's protests had obviously been encouraged by an address which T.A. Rickard had recently given to the Institute in British

Columbia on "The duties and privileges of technical journalism." Rickard, who a year earlier had sold his *Mining and Scientific Press* to McGraw-Hill for amalgamation with *E/MJ* in New York, had made a typically outspoken speech in which he had asserted that one of the reasons for his paper's disappearance was that the American Institute of Mining Engineers had started a monthly magazine to cover the same field. And he told the CIM: "You—the mining profession—are killing independent mining journalism for the sake of substituting something entirely different; you are developing a subsidized technical Press, controlled by a coterie."

The Journal followed up by renewing previous offers to put at the Institute's disposal a sufficient number of pages each week to publish its technical reports and any other material it wished to include. This prompted some lively discussion between George C. Mackenzie, on behalf of the Institute, and Mr. Harpell, for the Journal, all of which provided entertaining reading when published.

But the Institute would not be moved and *CIM Bulletin* continues to this day as *CMJ*'s chief competitor, and with a high reputation for its solid technological material, operational papers and articles covering the widest spectrum of Canada's mining interests which are of particular concern to its professional members.

CHAPTER EIGHT

A Very Worthy Doctor

Rossiter W. Raymond

Whilst Richard Rothwell will always remain the dominant figure in the history of the first 50 years of America's *Engineering and Mining Journal*, in the wider field of mining literature Raymond, his co-editor, was pre-eminent and, both by his pen and his personality, exerted a greater influence on the industry's development.

It is easy to describe a man of outstanding achievement by saying that he became "a legend in his own time". This was, however, undoubtedly true of Dr. Rossiter W. Raymond who at various times in an adventurous career was soldier, engineer, orator, editor, story-teller, poet, administrator, Biblical critic and theologian, teacher and chess player. In several of these capacities he was superlative. In none was he mediocre.

Walter Ingalls, who became editor of *E/MJ* in 1905 after sporadic association with the Journal over 15 years or more, has left a record of his first meeting with Raymond in 1889 at Leadville, where Ingalls was then a young mining engineer. "The honour of an introduction to this famous man overwhelmed me," he wrote "and I well remember the awe that I felt in entering his private car. It seemed to me that during our subsequent acquaintance and later friendship of nearly 30 years, he never changed greatly in appearance from that day. He never appeared to grow very old in looks, and certainly not in spirit. At the time of my first meeting with him he was in the midst of a game of chess, and I do not think he relished the interruption, although he pretended it did not matter. However, it was my good fortune soon to come close to him when, late in 1890, I came to New York to assist Rothwell on *E/MJ* and the latter, departing for a prolonged journey, enjoined upon me to look to Raymond as a guide, philosopher and friend, and until I left the paper we had such an association."

Ingalls did, in fact, become Raymond's chief biographer and his recollections of their association are the chief source from which his profile can be sketched today.

Rossiter Worthington Raymond was born in Cincinnati in 1840, son of Robert Raikes Raymond, whose ancestors came from England to Salem, Massachusetts, about the year 1632. His mother, Mary Ann Pratt before her marriage, was also from an old New England family. Eldest of seven children, Rossiter received his early education in the public schools of Syracuse, N.Y., and, in 1857, he entered the Brooklyn Polytechnic Institute, of which his uncle, Dr. John Raymond, was then the principal. After he had graduated at the head of his class, his parents decided to send him to Europe to continue his studies and he took passage in the clipper ship, *Great Western*, of the Black Ball line. It was a disastrous voyage. Crippled by gales and with half a rig, half a crew, and on half allowance of water, the ship finally crawled up the Irish Channel to Liverpool, to the amazement of her underwriters and to the relief of her passengers. Raymond, still only 18, had answered a call for volunteers to assist the depleted crew—and finished the voyage as third mate.

In Europe, Raymond headed for Germany, where he studied at Heidelburg and Munich, and then at the Mining Academy of Freiberg. Back in the United States, he joined the Federal Army in 1861, serving in the Civil War as aide-de-camp, with the rank of captain, on the staff of Major-General Frémont, the controversial explorer, politician and adventurer. During the campaign in the Valley of Virginia he was commended for gallant and meritorious conduct.

War over, Raymond looked around for means to put his Freiberg training into practice and to begin his career as a mining engineer. Whether he formed his association with Dr. Adelberg at once, or a little while later, is not clear, but his partnership with the accomplished chemist and technologist flourished, being almost coincident with the birth of the modern mining industry in the United States. It was on

95

Adelberg's advice that Raymond undertook the translation of Professor Bernard Cotta's *Die Geologie der Gegenwart* and offered it to the Journal—"partly", as Raymond explained later, "to help that enterprise along and partly—perhaps chiefly—to advertise our firm".

But the firm did not need a great deal of advertising. "Those were the days", wrote Raymond, "when mining experts were scarce and were eagerly employed. For some years, Adelberg and Raymond did an immense business. Herman Credner (afterwards professor of geology at the University of Leipzig and director of the Saxon Geological Survey), Anton Eilers (who became an eminent metallurgical authority and is well known for his part in the development of the American Smelting and Refining Company), Charles A. Stetefeldt, O.H. Hahn and others who subsequently distinguished themselves in American mining and metallurgy, were first employed as assistants by the firm. For some time my personal work was largely that of superintending the labours and editing the reports of such men."

Raymond's appointment as U.S. Commissioner of Mining Statistics followed shortly after his acceptance of the part-time job of editing the Journal, and for the next few years this occupied most of his time. He was always ready to acknowledge that it was not until Rothwell took the business of the paper in hand that it began to be profitable and to occupy the position in the industry which he had cherished for it but, for lack of time and money, he had been unable to achieve.

One of the chief products of Raymond's work as U.S. Commissioner of Mining Statistics was a series of eight volumes of detailed analysis of the mineral resources west of the Rocky Mountains. Ingalls had occasion to study them minutely in the course of a later investigation and was full of praise for "the beauty of their style; their clear language, incisive and easily flowing; the accuracy of the descriptions; the keenness of the perception manifested throughout; and the remarkable prophesies which had to wait only a few years for their fulfilment. Written 40 years ago when the Pacific railways had just been completed, and the Great American Desert was receiving the first strokes of development, especially as to its mining industry, these volumes can be read with almost as much interest today (1919) as when they were penned, and they may be read also with profit. No other country possesses such a chronicle of the development of its mining industry, and we may rightly esteem these volumes as classics of mining literature."

They were, however, written under some duress, and certainly the surveys had to be made on a very tight budget. Raymond himself recalled: "When I undertook the work I was hampered, on one hand, by youth, inexperience and the lack of means, and aided, on the other hand, by the cordial assistance of my old Freiberg colleagues, and the correspondents of the Journal of which I was editor. If I remember correctly, I had in the first year $5,000, and in no succeeding year of the eight during which I served, more than $15,000, to pay all salaries, travelling expenses, clerical and other labour involved in my work." There was obviously little to be saved out of his government earnings to keep the Journal going and it is no wonder that he had to sit up late at night writing children's stories to pay the printer. Moreover, as a Federal Government agent, he ran into much suspicion and distrust during his journeys in the West. The miners on the Pacific coast had settled their own laws and customs and did not relish Federal intervention.

Raymond did not overdraw the activity of this period of his life. In addition to the two responsible positions which he was filling, as well as his private professional work and his interest in the Institute, he accepted an appointment as lecturer on economic geology at Lafayette College, and occupied that chair until 1882, giving in one of those years the entire course on mining and metallurgy. In 1893, he was appointed U.S. Commissioner to the Vienna Exposition and there delivered an address in German on Patent Law, followed by a speech to an international meeting of geologists; and later, at Liège, in Belgium, a lecture to the Iron and Steel Institute.

From 1875 to 1895, Raymond was consulting engineer for a number of prominent iron and steel companies and coal mines in Pennsylvania. As president of the Alliance Coal Company, and through a personal friendship with Franklin B. Gowen, who was then the industrial monarch of the Pennsylvanian coalfields, he became concerned in the campaign against the "Molly Maguires", the secret society begun by

Irish immigrant mineworkers who resorted to vandalism, terror and murder in their struggle for better pay and working conditions.

There is no doubt that some of the collieries in the United States at this time were as bad as those in Britain before Lord Ashley's reforms and the improvements that Henry English and the *Mining Journal* had helped to bring about. Whatever Raymond may have thought about the validity of the miners' grievances, he was a vigorous opponent of all forms of tyranny carried out in the name of labour. His articles, "Labour and Law", "Labour and Liberty", and others published in the Journal at the time of the Homestead riots, attracted wide attention and for these, as well as for similarly straight-talking about the conduct of the Western Federation of Miners in Montana, Idaho and Colorado, he was denounced by the unions which he criticised. He was threatened with hanging if he should dare show his face at Butte, Montana, but he was not scared by such warnings and made a point of going to Butte at the first opportunity.

In the midst of all these preoccupations he still found time to undertake the direction of the *Saturday Evening Post*'s Popular Lectures on Science, which marked the beginning of what was to become a vast lecture system in the City of New York. He was also one of three New York State Commissioners of Electric Subways for Brooklyn, and served as secretary of the board, preparing its final report, which was accepted as a masterly statement of the problem of municipal engineering and policy involved in the distribution of electric conductors. The president of one of the companies concerned proclaimed him "one of the first American electrical subway engineers", but by this time Raymond had turned his attention to the development of the telephone, becoming consulting engineer to the New York and New Jersey Telephone Company.

Raymond's work for the American Institute of Mining Engineers between 1884 and 1912 has already been dealt with in the preceding chapter. But in the midst of all the pressures which this appointment added to an already extraordinary work-load, he was in 1898 admitted to the bar of the Supreme Court of New York, and of the Federal District and Circuit Courts, his practice being concentrated on cases involving either mining or patent law.

From his early years in the field, Raymond had always displayed a special interest in the science of economic geology, particularly ore deposits. His association with Credner in the old days of Adelberg and Raymond had encouraged this interest, and during his commissionership on the Pacific Coast he had devoted special attention to the nature and form of mineral deposits. Moreover, his mind was naturally analytical and deductive—an essentially legal mind. In the framing of its laws for national lands, the United States Government had not shown much comprehension of the peculiarities of the manner in which Nature had deposited valuable ores. The location of mining claims was largely governed by what became known as the "Law of the Apex", under which there was much room for dispute among claimants for the ownership of the same deposit, or part of it. Obviously, it became necessary to study, classify and interpret the ore deposit to ascertain if or how the law applied to it; this demanded the services of experts and it was natural that Raymond should be promptly requisitioned.

The first major case in which he appeared was the Eureka-Richmond, which was to become a classic in mining legal history. In summarizing Raymond's part in this case, William Scallon, one of his legal colleagues, wrote:

> The suit involved questions of fundamental importance, the correct determination of which was vital to the successful operation of the United States Mining Statute. The decision rendered in it has been ever since a beacon light for lawyers and for courts. It was Judge Field who rendered the decision, but it may well be said that it was Dr. Raymond who induced it, and the great jurist himself gave credit to him for his convincing exposition of the views which the court adopted; indeed, Judge Field adopted also his very words in the statement of these views. Dr. Raymond successfully maintained that the law was made for the practical miner and the prospector, rather than for the geologist, and should be read in the light of practical experience and interpreted as practical mining men would understand it.
>
> The law is, even now, the occasion of so many uncertainties that one can hardly imagine what condition would have ensued if Dr. Raymond had not, at that early day, succeeded in his advocacy of this beneficent rule. In the case referred to, it was as a witness that Dr.

Raymond advanced his views; since then he has published many
writings of the subject of the mining law. They have been no less
influential in assisting and directing legal and judicial thought.

Raymond's brilliant exposition in the Eureka-Richmond affair generated a great demand for his legal services. He was invariably on the
winning side; once, his opponents surrendered before trial; on another
occasion, though not at that time a member of the Bar, he was invited
by the U.S. Supreme Court to address it on a point of mining law and
his exposition was accepted by the court in rendering its judgment.
However, while an expert in its interpretation, he deprecated the law
which fostered such litigation. In an article on "The Law of the Apex",
he exposed its stupidity and dangers, and he made it part of his life-
work to fight for its amendment.

But of all his accomplishments, it is his flair for the written word
for which Rossiter Raymond will be longest remembered. His style of
technical writing was as near perfect as anything could be. According
to Walter Ingalls, he was "the supreme master of us all". During
Ingall's editorship he made a standing rule as far as Raymond's
contributions were concerned: if they were to go to the composing
room at all—and most of them did—they were to go exactly as he
penned them. "The characteristics of his style", said Ingalls, "were
incisiveness, lucidity and a wonderful fluency. He had the gift of
choosing exactly the right word, for constructing ingenious and engaging phrases, for producing well-balanced sentences, whilst his thoughts
were developed in a logical array of consequential paragraphs."

Nearly all of Raymond's articles for the Journal, which were
continued to the year of his death, and all his private letters, were
written by his own hand. He did not like to dictate, and perhaps the
excellence of his literary work had some relevance to this, for there
are few indeed, even today, who can produce a flowing style by
mechanical means. His penmanship was graceful and easily legible.
Interlineations and corrections were uncommon. To compositors and
to proof readers, his "copy" was always a joy.

But Raymond was not without his faults and foibles. It was not
surprising that a man so steadfast in his convictions should sometimes
be opinionated and domineering and occasionally display considerable

egotism. These traits often caused his younger associates to fear lest they should offend him. However, Raymond seldom took offence, even if it were a refusal of himself. This was put to the test quite early in Walter Ingalls's association with the man who was regarded with such awe among the Journal's editorial staff. "Soon after I had become editor", recalled Ingalls, "he sent me his best wishes in a very warm letter and at the same time proposed a series of articles on a certain topic, which I accepted in principle."

"However, when I received his manuscript, I was considerably exercised over his treatment of the subject and was much disturbed over how to tell him just what I thought. I made up my mind that the only thing was to be perfectly frank, so I wrote him that I did not think it wise either for the Journal, or for himself, to publish the articles, and, in short, I rejected them. To have to reject an article by Dr. Raymond was an appalling thought to me in my newness in the editorial chair, and I awaited his reply with a good deal of trepidation, really expecting something tart. I was immensely relieved when he wrote and told me that I had done right. The flow of his usual contributions on his own good subjects kept right on."

Throughout Raymond's career, many honours were inevitably showered upon him. He was made an honorary member of the American Philosophical Society, the Society of Civil Engineers of France, the Iron and Steel Institute, the Canadian Mining Institute, the Mining Society of Nova Scotia, the Institution of Mining and Metallurgy in Britain, and numerous other technical and scientific organizations both at home and abroad. He received the degree of Ph.D. from Lafayette College and that of LL.D from Lehigh University. On the latter occasion, in 1906, speaking as an adopted alumnus of the university, he delivered to the graduating class an address on "Professional Ethics" which became a standard lecture. In 1908, the Institution of Mining and Metallurgy awarded him its gold medal. During a visit to Japan of a party of members of the American Institute of Mining Engineers, Raymond received from the Mikado the distinction of Chevalier of the Order of the Rising Sun, the highest order ever given to foreigners not of royal blood, "for eminent services to the mining industry of Japan". These services consisted of advice and assistance rendered in America to Japanese engineers, students and officials over a period of more than 25 years. In February 1915, Raymond delivered the commemorative address on the 150th anniversary of the foundation of the University of Pittsburgh, and received its honorary degree of LL.D.

To this list of honours, many more could be added, but Raymond also had a very full life outside his professional interests. From his earliest years he was a pillar of the Plymouth (Congregational) Church in Brooklyn, and close friend of its controversial founder, Henry Ward Beecher, regarded by some as "the greatest preacher since St. Paul preached on Mars Hill". A son of Lyman Beecher, whose family of 13 brilliant children included Harriet Beecher Stowe, author of *Uncle Tom's Cabin*, which was to stir the World's conscience on the Negro question, Beecher's oratory drew weekly congregations of 2,500, and his church claimed the largest membership in the United States. Raymond took charge of the Sunday School, initiating a custom, carried on for 50 years, of writing an annual story, which he read to the children at Christmas time. He never failed to make this contribution and the stories were subsequently printed for the benefit of his scholars and for his friends outside the church circle.

When Beecher had to leave in the middle of a service one evening in 1887 to keep an engagement elsewhere, he called Raymond to the platform to take charge and, turning to the congregation, said "You see, I leave you in good hands." The remark might have been prophetic, for Beecher never returned; he was stricken with apoplexy and died. After his funeral, church leaders went to Raymond and asked him if he would give up his career as editor, engineer and lawyer to take on the pastorate. This invitation, which was not generally known until some years later, was not altogether surprising as Raymond had gone deeper into theology than most clergymen. During the winter of 1885, as Beecher preached in the auditorium of the church, Raymond had lectured on the history of Israel and Old Testament problems. This was the time when the Presbyterian Church was in the throes of its heresy trial of the Old Testament theologian, Professor Charles Briggs. Whilst other churches were torn apart by the controversy, the Plymouth Church, due in no small measure to Dr. Raymond, went through this difficult period without disturbance, upheaval or fear. However, Raymond declined the invitation; he felt that Providence had guided him into an appointed path and, with his life's journey two-thirds filled, he did not think that it was the Lord's wish that he should now reverse that path.

Raymond found one of his few relaxations in chess, of which he was a skilful player, and when, on his 70th birthday, he was presented by James Douglas, on behalf of his friends, with a silver service, he was delighted to find that one of the pieces was engraved with the symbols of the game. In making the presentation, Douglas recalled that Raymond had once drawn a hard-fought game with Wilhelm Steinitz, one of the masters of the day; he also referred to an occasion in 1908 when "a confident group of chess players on the *Campania* challenged players on the *Oceanic* to a game by wireless telegraph, and got the worst of it because Dr. Raymond directed the defence".

Raymond died suddenly in the afternoon of the last day of 1918. He had not been well for a few days, but the trouble did not seem to be of much consequence. On the previous Sunday he read his 50th annual Christmas story to the Sunday School and the same week-end he gave a long address on the spiritual influence of the World War. He had been asked to speak on the forgiveness of enemies, but he was not altogether keen on this theme, believing very deeply in the holiness of the Allied cause, and abhoring particularly the treatment of the Belgians by the German armies. In fact, he told his audience, he found himself in sympathy with some of the imprecatory psalms, and that he well understood the reaction of the South Seas Island cannibal who, when asked if he had forgiven his enemies, replied "Certainly, certainly; last week I ate them all up." But he then went on to give a brilliant address on what the war had done to the soul of the American people. "Never was he in finer form", wrote one of his friends. "Notwithstanding his 79 years, his mind glowed and sparkled through 40 minutes. Never did he seem more lovable, more gentle. There was in his carriage always a certain note of chivalry. He seemed a 20th century embodiment of the Knights Errant of Sir Thomas Mallory. Those who recall his youth think of him as a kind of Sir

Galahad, and those who remember his age think of him as Sir Percival, who saw the Holy Grail because his heart was pure."

In one year, the American mining industry lost three great men—Anton Eilers, James Douglas, and Rossiter Raymond. They were born at about the same time and died within the same narrow space. Together they probably did more for American mining than any other three men during a similar period of time. They were all great friends, the friendship between Raymond and Eilers being perhaps the most profound. Eilers once said that in more than half a century he could not recall a single occasion of even temporary discord to mar their relationship. At Eilers's funeral, Raymond's oration was not only a moving tribute to the memory of his friend, but an exaltation of his own religious philosophy:

> And yet, can this be the end? Love says, No! and science says, No! for the very existence of science demands the assumption of a rational universe. Nature must not tell us lies. Her evidence must be interpretable. And science, starting on that basis of faith in honest evidence, has discovered not only order, but purpose in the evolution of the universe. We can read the progress of that purpose in ages past, from primordial slime through the ascending forms of life, to savage man, barbaric man, civilized man—through specific features to personality.

> Science confirms the poet who says 'An honest man's the noblest work of God.' And science cannot receive with respect the conception of a Being who would spend aeons in careful, patient preparation to bring forth a man only to destroy him—a God blowing bubbles and, just when such a radiant sphere has reached its brightest rainbow glory, dashing it into mist, in order to begin another with futile inflation. How childish, how absurd! No, we demand a reasonable universe and a respectable God—we students of science. We will not accept the notion that earthly death ends all. It is too ridiculous.

Almost any man can be deemed distinguished if such can be said of him in any one major accomplishment. It was given to Raymond to achieve eminence in several fields, but predominantly in two—as an editor and publicist, and as a professional specialist in one of the world's oldest industries during a period of rapid technological advance.

In Raymond's own view, however, he had never deliberately set out to achieve any specific targets, nor to fulfil any one ambition. Rather did he unconsciously obey the injunction, "Get thy tools ready; God will find thee work."

God found him much to do, and he did it well.

CHAPTER NINE

Melbourne 1840 – before the gold rush

Australia and S. Africa

The predominant position which New South Wales occupied at the Mining Exhibition at the Crystal Palace in 1890, and the great interest taken by its government representatives in getting the Institution of Mining and Metallurgy started a couple of years later, reflected the rapid progress of mining in this, and other, Australian colonies in the second half of the 19th century.

The discovery of gold in Australia in 1851 not only marked the opening up of the continent, but helped to solve a serious problem of unemployment in Britain, where London and other cities were suffering acutely from over-population. If any man can be said to have provided the answer to this problem, however unwittingly, he is Edward Hammond Hargreaves. A year or two earlier, he had left his sheep farm near Sydney to join the California gold rush. There, he had been struck by the similarity of the local geological features to an area across the Blue Mountains, 100 miles or so north-west of Sydney. Returning home, he almost immediately found gold at a place now named Ophir, and subsequently also, about two miles from this spot, at Summer Hill.

Not wishing to be outdone, the authorities in Melbourne organized a Gold Discovery Committee and offered £200 for proof of the presence of gold within 200 miles of the settlement. James William Esmond who, like Hargreaves, had gone to California to try his luck, also noticed geological occurrences resembling those near his own home at Clunes, and he set off to put his theories to the test. By an

106

extraordinary coincidence, Esmond and Hargreaves sailed in the same ship from California back to Australia. There is no record of what they talked about on the long voyage—or indeed whether they talked to each other at all—but Hargreaves discovered his gold at Ophir on February 12, 1851, and on July 1, the day on which the Melbourne settlement was proclaimed a separate colony and named Victoria in honour of the Queen, Esmond found his gold near Clunes, encrusted though it was with obstinate quartz.

It was California all over again. The Rev. James Ballantyne, of Melbourne, in his book *Homes and Homesteads of Victoria* wrote that the discoveries "changed, as by a wave of the magician's wand, every feature of life in Australia. The pulse of the community, which erstwhile beat quietly and steadily, at once mounted to fever-heat. There was but one theme on every lip, and that theme was 'gold'. It intoxicated the whole body of the people. They rushed pell-mell to the various spots where the dazzling metal was supposed to be obtainable. The labourer left his implements of toil and ran; the mechanic quitted his bench; the clerk immediately threw up his situation; the merchant left his counting-room; the barrister left his case unfinished. Melbourne was all but deserted. In the course of a few months about one-half of the male population of the colony had left their wonted avocations and gone on the popular adventure. Then, too, the people came in hot haste from the neighbouring colonies, crowd following crowd as fast as ships by sea and conveyances by land would bring them—men of every shade of character, and thousands with no character at all, each and every one attracted by the bewildering glare of virgin gold. Little wonder that business came to a standstill, that the old landmarks were torn up, that the foundations of society were out of course."

Within a year there were nearly a quarter of a million immigrants at the Victoria diggings. By 1856, the population of the colony was five times that of 1851; the yield of gold, valued at £13 million in 1852, rose to more than £100 million in 1860. "Thus," wrote one historian, "the small colony of Victoria was, as the apt phrase goes, precipitated into a nation."

In New South Wales, the picture was very much the same. Here, the first alluvial diggings were the richest, but never so alluring as the

Victorian fields. For the first 20 years or so, gold mining remained almost exclusively a small-working operation; nevertheless, the industry put new life into the wastelands, and Sydney blossomed into a city which promised to become mistress of the southern seas.

This was the heroic period of Australian mining, an adventure which took prospectors deeper and deeper into remote and hostile territory, treading where no white man had hitherto ventured, penetrating as far as the Palmer River, in North Queensland; to Halls Creek, in the Kimberley region of Western Australia; and culminating in the 1890s with the discovery of Coolgardie and Kalgoorlie.

Whilst these events resulted in a temporary redistribution of population, and in some areas—as in South Australia—cities were left desolate and deserted, the enormous influx of people into the colonies from overseas was of great permanent benefit. And in Britain, which provided most of the immigrants, although at first the depletion of the labour market caused some embarrassment—especially as there was a great demand for British goods in those Australian colonies to which the people had flocked—the effect in the longer term was to give an impetus to industry and a considerable rise in the value of skilled labour.

As the colony which had been "precipitated into a nation" almost overnight by its mining, it was appropriate that Victoria should produce the first Australian paper devoted entirely to mining affairs. The *Australian Mining Standard* was founded in Melbourne in 1888, by which time a stable industry was beginning to take shape out of the rash of individual ventures and co-operatives. Unfortunately, hardly any records remain today of this paper despite the fact that it continued publication for more than 70 years, eventually changing its character and its name in 1960, when it claimed to become the first business and investment weekly magazine in Australia. Alas, it chose a bad time for such metamorphosis and it died a pathetic death the following year, lamenting that "Australia's economic circumstances have brought investment to a lower level than many years".

Australia's oldest mining paper still in production is the *Queensland Government Mining Journal*, which was started in 1900 when the colony was in the midst of its early mining boom. Both Victoria and

New South Wales were producing vast quantities of gold before the Moreton Bay settlement, centred on what is now Brisbane, was hived off from New South Wales in 1859 to form the nucleus of the new colony of Queensland. Its mineral wealth was at this time almost totally unexplored—understandably, as in addition to the enormous size of the territory and its sparse white population, there were dense tropical scrublands to be negotiated and, worst of all, large numbers of unfriendly natives, who had already resisted with their spears the advance of shepherds and stockmen in search of more room for their increasing herds. However, the pastoralists pushed on, planting sugar and laying the foundations of a big new industry, braving climate, terrain and massacre, penetrating into Cape York, and to the furthest northern limits of Australia.

Railways and other extensive public works were at this time being undertaken to open up the country, but suddenly the crash came. The London bank, with which the government had an agreement for funds, failed; credit dried up and businesses, institutions, societies and traders all went down before the financial storm of 1866. There was widespread unemployment, including many hundreds of men on construction work on the railways. The position of the young colony was desperate.

Then, one evening in 1867, John Nash, an out-of-work miner, camped on the track to Maryborough, at a place called Gympie, and, while brewing up for tea, found gold among the quartz lodes near at hand. From then on, discovery followed discovery—the alluvial fields at Cape River (1867), Ravenswood (1868), Gilbert (1869) and Etheridge (1870). Then, in 1872, the important lode mine of Charters Towers was found, an industry was established and a colourful community developed into the second city of the colony. Queensland was saved by gold—as it has been several times since in times of crisis.

The establishment of the *Queensland Government Mining Journal* was a remarkable piece of foresight by the Department of Mines at a time when few governments in the world had shown much concern for promoting information about the mining industry, or about any other national industry for that matter. It has perhaps some parallel with the decision of the *Comité de Salut Public* to begin publishing *Annales des Mines* during the French Revolution, though the immediate objectives were somewhat different. The French paper was started very much as a political organ to prove to the proletariat that the new government was going to run mining in the interests of the many and not for the privileged few, and only slowly did it develop its high quality of technological reporting and commentary. The Queensland magazine, on the other hand, was launched positively for the benefit of the industry in the colony and, more especially, to assist mining men of all grades to become more productive and conscious of the need to work safely. It, too, developed its technical content gradually as time went on.

However, readers of the editorial in the first issue of the *QGMJ* might well have wondered what was in store for them. In addition to the traditional statement "eschewing politics", and a declaration of intent to open the paper's columns to correspondence which "must be devoid of personality and couched in concise form", the editor assured Queensland's miners that "poetry will not be excluded, but its numbers and rhythm must be worthy of the Genesis it sings". There was also a warning to potential advertisers that "no advertisements will be accepted of an offensive, indelicate or charlatanic description".

The article, in fact, reflected more the character and background of the editor than the policy of the paper. The first incumbent, W.O. Hodgkinson, apart from being a most erudite gentleman, had already had a varied career in which he had been a midshipman in the Merchant Marine, a member of the War Office staff in London, explorer, prospector, Mining Warden and Minister of the Crown in Queensland. He played a prominent part in opening up the Palmer River goldfields in the far north when the going was exceptionally rugged. To encourage poetic contributions, he penned some verses of his own for the first issue of the Journal, and was also responsible for a cover design, emblematic of the mining industry and including many of the metals and gems known to occur in the colony.

Some of his garrulity reached unusual heights—or depths—even for these days of literary extravagance. "It is," he wrote, "from the thousand rills of information trickling through a thousand observant men that must issue the full stream of data with which it is desired to irrigate the public mind for, be it understood, the material welfare of the community at large . . . We strike out on our new path buoyed by the belief that the need of an authoritative exponent of mining, as she is and ought to be, is patent to a wider sphere than our local world, and that such need may be met by the display of an energetic, patient determination to traverse the path leading to success."

Alas, the path that Mr. Hodgkinson was to tread was but a short one. He died within a couple of months and was succeeded by Walter Morley, who had a sound background of journalistic experience on newspapers in Brisbane. Morley edited the Journal for 21 years and gave it the stability and professional standards that it needed. He was followed by another Australian newspaperman, Chester Reynolds, who held the post until 1947, being succeeded in turn by Claude Simmins until 1962, and Peter Wakeling until 1977. The present editor, Joe Colgan, is therefore only the fifth since Hodgkinson's brief occupancy of the chair in 1900.

Looking back through the files, Joe Colgan acknowledges that it would be difficult to attribute any major mining development in Queensland to *QGMJ*. "The Journal", he emphasizes, "was founded when there were thousands of small-time prospectors working in remote parts of the territory. The idea was to give these men official information, help and encouragement, and to tell them about prices, equipment, methods and developments in various districts. I think that the Journal helped to create and preserve a kind of brotherhood among mining people in Queensland, particularly in the lean years between the turn of the century and the Second World War when substantial new discoveries were few and yet there were pockets of prosperity and joy in a wilderness of back-breaking work, disappointment and frustration . . ."

Today, little or no gold is being recovered in Queensland, whose mining industry is based largely on coal, copper, lead, zinc, bauxite—and uranium. More than 70 per cent of Australian copper comes from Queensland's Mount Isa complex, which is also the largest single producer of lead and zinc in Australia, and the Mary Kathleen is, at the time of writing, the only Australian uranium producer, though the long wait for full-scale uranium development in the continent should come to an end when the question of royalty payments to the aboriginals and environmental regulations in the Northern Territory have been successfully resolved. The colourful, exciting days of the alluvial gold diggings are long past and, as in most parts of the world, mining is the concern of large companies and powerful financial consortia. Politically also, Queensland has undergone great and frequent change; it is remarkable that any paper produced by a government department should survive so many variations in policy and it is sufficient tribute to the *Queensland Government Mining Journal* that, after nearly 80 years, it is still serving the interests of an industry which plays such a major part in the economy.

*

One day in 1908 a young man and an office boy walked down Queen Street in Melbourne and talked about an idea. The man was Peter Tait, the boy was Victor Kinsman, and the matter they were discussing was the establishment of a paper which would cover the Australian mining scene as a whole.

Before the year was out the idea had become a reality: the first issue of the *Australian Mining and Engineering Review* appeared on October 5 with the imprint of the Tait Publishing Company, a name which was to become synonymous with Australian mining journalism for more than half a century. Tait died in 1953, and Kinsman about 10 years later—at about the time when the Tait companies became wholly-owned subsidiaries of Thomson Publications (Australia).

Today's *Australian Mining* is a direct descendant of Tait's first magazine and, despite several changes in title in the years between, it still retains something of the stamp of the man who was its first editor, advertising manager and proprietor. The first editorial expressed his ideas very clearly: it is obvious that he wrote it, though the paper was born and developed in the old tradition of editorial anonymity. "The most energetic centres of our scientific development are", he wrote, "situated in remote corners of the continent, and in consequence those engaged in the elucidation of nature's problems are to a large extent banished from the world known to city folk. To minimize the isolation imposed, this journal proposes to provide a reliable record of progress in mining and engineering science. Thus, we hope to become the medium of benefit between the mining man desirous of keeping abreast of the times, and the manufacturer who can provide him with the plant to realize this laudable object."

Unlike many other mining journals which were the product of boom conditions, the new Australian paper began life when there was something of a recession: "That the mining industry is at present experiencing a period of depression is an undeniable fact, but that this state of affairs is merely one of the recurring changes following on the swing of the pendulum between supply and demand, can also be confidently assumed ... All things considered, and after assimilation of the foregoing facts, it may be pointed out that the cloud at present shadowing the mining industry has indeed a silver lining."

Prophetic words. As the industry picked up, so the paper grew in stature and other publications were added to the Tait stable: first, the weekly *Tenders* in 1912; the *Commonwealth Engineer* (later *Australian Civil Engineering*) in 1913; *Electrical Engineer* in 1924; and *Manufacturing and Management* in 1946. In 1947, the name of the first-born was changed to *Chemical Engineering and Mining Review*. This title was retained until 1960, when the order of the words was re-arranged, to place the emphasis on "mining", though "chemical engineering" had always been related to minerals and the allied treatment processes, except for a period during the 1930s when the publication was the official organ of the Royal Australian Chemical Institute. The present title, *Australian Mining*, appeared with the January 1966 issue, the change being appropriate at a time when mining and minerals were having the greatest influence on the economy of the nation.

One interesting off-shoot of the Tait publications was a book company which was developed out of a service given to early readers in outback areas. As demand for publications and books grew, so also did the stocks around the office in anticipation of visits to Melbourne of prospectors and miners from the interior. The bookshop soon came to be recognized as an organization with a genuine policy of service, and became known to technical men throughout the Commonwealth.

Best-known among the editors who succeeded Tait were Harold Elford, Maurice McKeown, Angus Norrie and John M. Dew. All knew and worked with Peter Tait—and were influenced by him. Elford, a mining engineer, became managing director when Tait retired in 1952 and after his death the companies were bought by A.R. Everett, who subsequently negotiated their take-over by the Thomson Organization.

Whenever any international or multi-national group assumes control of a local enterprise which has been built around the personality and dedication of a few enthusiasts, there is always risk that the new organization will lose its character and direction of purpose. This is particularly true in the publishing business. Such fears must have been felt in Melbourne when Thomson's exercised an option to take over the Tait publications in 1964. But, in the words of John Dew, who retired as editor in 1969 when the group moved to Sydney: "We realized how closely the policy of the Tait companies was identified with that of the new owners when we read the views of Lord Thomson in his chairman's statement at the 41st annual general meeting." In that statement, Lord Thomson reaffirmed his group's determination to ensure that each of its widely-spread publications should serve to the highest possible degree the interests of the community or population by whom it is read.

*

If the impact of mining journalism on the industry in the Commonwealth of Australia has been relatively slight, it is little to be wondered at in view of the vastness of the territory and the fact that its mineral wealth has been found to be widely scattered over its dusty face.

In South Africa, less than one-sixth the size of the Commonwealth, and only half as big as Western Australia, its largest state, it is more difficult to understand why the media has had so little influence on the progress and discipline of the industry. It is even more surprising in view of the outstanding role that mining has played in a country whose industrial history is studded with such romantic names as Cecil Rhodes, C.D. Rudd, George Farrar, Abe Bailey, Barney Barnato,

114

Solly Joel, Alfred Beit, John Hays Hammond, Lionel Phillips, and in more recent times the Oppenheimers, father and son. Their story has, of course, since been told, with the benefit of hindsight, by a number of South African authors, including the late A.P. (Paddy) Cartwright whose book, *The Gold Miners*, is an outstanding example of how industry can be so sensitively treated as to produce a literary classic.

However, research in South Africa has failed to produce evidence of anything like the involvement which the Press has had in the development of the mining and metals industry in Europe and throughout North America. Perhaps this is because mining has been so central to the economy and such an integral part of everyday life that the daily Press has also been, in effect, the mining Press. However, even accepting that the sole function of the South African media has been to chronicle, rather than to attempt to influence, mining policy and development, its record is obscure, and in the offices of those few papers which have their roots in the heroic days there appears to be little knowledge or appreciation of the work of early journalists or journals. Perhaps events have moved too fast for a young country to absorb; in a materialistic world, the importance of the preservation of archives is often appreciated only when it is too late, and history becomes distorted in a tangle of argument and, alas, is often influenced by current political expediency.

From the days when the discovery of diamonds and gold began to transform the South African economy from subsistence agriculture to rapidly developing industrialization, the lives of so much of its population, both white and black, have been bound up with mining. In February 1886, George Harrison, an itinerant handyman and down-on-his-luck prospector, stumbled on a gold-bearing outcrop that later proved to be the "Main Reef" of the great Witwatersrand Geological Basin. Harrison, having made his discovery, disappeared into obscurity, and his place was taken by men from all over the world, lured to South Africa, as so many of them had already been to California and Australia, by the prospects of a golden fortune. A mining camp quickly grew up and, equally quickly, became the city of Johannesburg.

Its first newspaper, the *Diggers News and Witwatersrand Advertiser*, was started barely a year after Harrison's discovery, and it is said that it was first in the field because the printing press of its rival fell from an overturned ox-wagon into the Vaal River. It was backed by friends of President Paul Kruger, but ran into trouble and soon folded in the face of the mounting competition. By 1890, there were no fewer than seven daily newspapers being published in Johannesburg; of these, only *The Star* survived, to be joined in 1902 by the *Rand Daily Mail*, and in later years by *Die Transvaaler*, *Die Vaderland* and *Beeld*, the five dailies at present serving the Witwatersrand.

The *Rand Daily Mail* had a sensational start under the editorship of Edgar Wallace, whose imaginative brain also conceived *King Kong* and a host of other thrillers. As correspondent for the London *Daily Mail*, Wallace had obtained a world scoop with the news that the Vereeniging Treaty, ending the Boer War, had been signed. His South African editorship lasted only one year—a year in which he won the paper world renown but, by his extravagance, also brought it to the brink of disaster.

The Star has also had its great editors, among them William (Jock) Addison, who later became Speaker of the Rhodesian Legislative Assembly; Horace Flather; J.W. Patten; and Rene de Villiers. The paper was also probably the first in South Africa to appoint a professionally qualified mining man to its editorial staff to cover industry affairs. Dr. J.T. Carrick, a Scottish geologist, occupied this post during the first decade of the century, and he could well have become the first newspaperman to make a major contribution to the industry in South Africa. He believed strongly in the existence of a gold field in the Orange Free State—almost 40 years before it was proved. In 1909, he decided to float a company in London, and in August of that year he sailed from Durban with some 300 other passengers on board the *Waratah*. The ship was never heard of again.

The first, and for many years the only, paper devoted entirely to the interests of the industry was the *South African Mining and Engineering Journal*, which was founded in 1892 and, except for its closure during the Anglo-Boer War, has appeared regularly ever since. For the first 78 years of its life it was a lively weekly, giving a mixture of technical and financial news as well as detailed technical articles, and for many years it also incorporated the *Johannesburg Stock Exchange*

Official Gazette. In 1970, it changed to a monthly in order to concentrate more effectively on technical coverage of mining engineering, and six months later, together with its Year Book, it became part of Lord Thomson's publishing empire.

In an interesting statement of policy following its reappearance after the Boer War, the Journal declared its independence of "any financial house, its representative or agent", and added: "Everything that makes for sound economics and improved methods of mining will have our warmest support, but these must be conditioned by whatever measures are necessary to uphold the prestige of the white races and to encourage the expansion of profitable work in every direction, commercial, agricultural and industrial. We hope to maintain an impartial course, dealing justly by labourer and employer, consumer and purveyor."

The big issue of the day was how best to get the mines going again after the war. By the time the Treaty of Vereeniging had been signed the economy of the Transvaal lay in ruins; the whole country had gone back 40 years. As Paddy Cartwright put it in *The Gold Miners*: "Eager as the mining companies were to get their batteries going again, their hands were tied by lack of labour. Managers, miners, administrative staff were scattered throughout South Africa. The native labour force that had numbered 110,000 before the war had vanished, nor were the 'mine boys' showing any anxiety to return to their jobs. The white man had been fighting and might fight again. It would be wise to stay in one's kraal until it had all died down. The British were said to have won the war, but who knew? 'Let us wait and see what happens', said the headmen. 'Time will show.'"

And pay was a factor, too. Before the war the mines had paid an average of £2:10s. a month with food and lodging. This was considered a high rate for native workers in those days; indeed, a carefully considered article in the *Mining and Engineering Journal* in its first year of publication had explained why £1 a month was adequate for a black mineworker. The Boers certainly thought the Africans were over-paid and when they took over at the beginning of the war they laid down a flat rate of £1. As soon as the Chamber of Mines was re-established, it increased this to £1.10s. But the Africans remembered the £2.10s. of pre-war days. Besides, there was plenty of alternative work available on the railways and various other reconstruction schemes that Lord Milner, the British High Commissioner, had put into operation.

The problem was critical. Only one thing stood between the mining companies and the promise of immense profits for the next 15 or 20 years—the shortage of labour. Thoughts began to turn to the possibility of importing unskilled workers from China. At first, the *Mining and Engineering Journal* was opposed to the idea: "We shall insist upon the principle that the ordinary labour supply of South Africa must be called upon to furnish South Africa's labour needs before experiments of introducing foreign labourers can be permitted," said one article in 1903. And again: "The Kaffir can very well do without us, as he did before—according to his lights. And we could do without the Kaffir if we only were without him. But he is here, and we have got to make the best of him, if only for our own sakes. If there were no mines at all, the native question would still be the greatest problem before the people of South Africa: our morals preclude us from the civilizing methods adopted towards the aborigines of North America and Australia, and our safety, if not our righteousness, forbid the use of overproof rum. . . . Our duty to posterity is clear and cogent: we may not, through neglect, leave a native problem so stupendous and unwieldy as to be overcome, if at all, by explosives only."

Another early opponent to the importation of foreign labour was the then-editor of *The Star*, William Monypenny; his directors disagreed with him and he resigned as a matter of principle. Meanwhile, public opinion became more and more in favour of looking overseas for the solution; it was remembered that even before the war much of the mine labour had been recruited from outside (notably from Mozambique) and, as the mines went deeper and deeper, there had been talk of bringing in indentured labour from other countries.

Finally, even Lord Milner, who wanted all his settlers to be white and British, decided to recommend a "crash" policy of allowing imported Chinese labour to fill the gap. As Professor Walker, the historian, put it: "The Chinese were to be the scaffolding which could be dispensed with as soon as a great white society had arisen within it, freed from dependence on Mozambique and the tribes."

The impact of 50,000 coolies proved not as socially disastrous as the opponents of the scheme had feared. They marched to the mines in their thousands—and then disappeared from public view behind the gates. However, the "Chinese peril" continued to worry the people of the Transvaal. In Britain, some horrific, and mostly untrue, stories of their treatment became a political issue which was a major factor in the fall of the Balfour government and the return of Campbell-Bannerman's Liberals with an impressive majority. The new political climate at home also encouraged the rise of the *Het Volk* party in the Transvaal and increased the influence of the Boer General Smuts, who was to become the first Prime Minister of a united South Africa.

And what of the Chinese? For the most part law-abiding and hard-working, they put the South African mines back on their feet, and five years after their arrival they were shipped back to China, profoundly ignorant of what they had accomplished. Subsequent developments in South Africa, both in the political and economic fields, have had little effect on the *Mining and Engineering Journal*—and the Journal appears to have had little effect on the industry. Unfortunately, the problem of employment in the mines and the whole question of social and industrial relationship between the races are today no nearer being settled than they were 75 years ago when the Journal pleaded that such issues should not be handed down to posterity to face a solution which would inevitably be attended by violence. The problems have since assumed a magnitude which finds almost daily repercussion in international councils. The mining Press can have but a small voice in debating these issues, which are at the very focal point of South Africa's destiny, but the newspapers of the world's great powers, and the international media generally, are never shy of expressing their views on a political, social and industrial system which divides humanity into classes according to colour of skin and which condones so many indignities. However, as the Journal observed in 1903, if there were no mines at all, these issues would still be the greatest problem of the people of South Africa. Accordingly, the Journal maintains a professional aloofness to matters which it does not regard as its specific concern, and its present publishers and editors see their function simply as a communications team serving the industry in the best way possible. For the past 25 years it has not had the technical field to itself, being joined in Johannesburg by the Pithead Press, which publishes a monthly magazine under the somewhat unwieldy title *Coal, Gold and Base Minerals of Southern Africa*, and whose circulation is about the same as the Journal's 4,000 copies an issue.

However, the combined circulation of these two papers is but a fraction of that of *Mining Survey*, which for the past 30 years has been issued by the Chamber of Mines of South Africa. Originally designed to encourage greater local recognition of the common interests of the mining industry and the nation, it has today a much sharper international focus. It publishes authoritative articles about the industry and international markets, and deals with a variety of other subjects, including investment opportunities and industrial policy matters. At present, *Mining Survey* prints about 34,000 copies in English and 3,000 in Afrikaans, and outside South Africa its circulation extends to people with mining interests in Britain, North and South America, India, Australia, Europe (including Eastern Europe) and various parts of Africa.

CHAPTER TEN

Competition in London

Edward Baliol Scott

Towards the end of the 19th century, Captain Laurence Hartshorne Scott, of the 11th Regiment (Devonshires), former Adjutant to Sir William Dennison, Governor of New South Wales, died in Australia. At the time there was nothing to connect his death with mining or mining journalism but, in the event, it was to have a significant impact on the development of Britain's leading publication in this field.

Captain Scott was the head of an ancient family in the United Kingdom, Scott of Scot's Hall, at Smeeth in Kent, which traces its descent from John de Balliol and his wife Devorgilda, founders of Balliol College, Oxford. John de Balliol's son John and grandson Edward were both Kings of Scotland—"Norman interlopers put into Scotland by Edward I to keep the natives quiet", as the present head of the family, Ursel Baliol Scott, rather irreverently puts it. But the natives did not remain quiet and survivors of the family eventually settled in Kent.

Whilst serving in Australia, Captain Scott took as his bride Rebecca Bettington, grand-daughter of William Lawton, pioneer explorer and one of a party of three who were the first to penetrate the Blue Mountains on an expedition which marked the beginning of the occupation of the interior—and the beginning of the problems of its more extensive exploration.

After her husband's death, Rebecca came to live in Britain in 1886. Some years later, seeking an investment for her money, she was advised to buy into the *Mining Journal* and, in 1898, she purchased, in the name of her son Edward, the company's mortgage and acquired part ownership from the mortgagee, Richard John Middleton, son of Richard Thomas (see page 64). The mortgagor had been the late Joseph McCrae, from whose executors Edward subsequently acquired the remainder of the company's shares.

In the early days of Baliol Scott ownership it looked as if Rebecca had not been very well advised for, despite the outstanding success of the International Mining Exhibition, which the paper had been largely responsible for organizing, and the prestige it had acquired for its part in launching the Institution of Mining and Metallurgy, the Journal, at the turn of the century, was going through an unsettled period of management problems which at times threatened its survival.

The solution of these problems devolved upon Edward. The only son of the marriage, and now in his late 20s, he had attended Geelong Grammar School before coming "home" in 1886 to go to Bath College and University College, Oxford, where he took an honours degree in Greats. He was called to the Bar and joined the Oxford Circuit, practising until the increasing demands of his mother's interest in the *Mining Journal* obliged him to discontinue what promised to be a highly successful legal career. He took charge of the Journal in 1901 and assumed the editorial chair, which he continued to occupy for the next 53 years.

In retrospect, it may be considered that Edward tended to concentrate more on the editorial side at the expense of the economic aspects of the business, but this was not surprising for anyone of his background. A classicist and a man of wide interests, his approach to the mining industry and its problems was, in the best sense, amateur without ever being amateurish. With none of the advantages or limitations of a specialist science training, he retained throughout his career a detachment and breadth of view which gave the Journal under his editorship an independence and an outspokenness rare among publications devoted to sectional interests. Inevitably, these qualities did not always contribute to his popularity in the mining industry—not least when, with annoying frequency, his reading of a situation turned out to be right and professional experts were wrong. Indeed, his piercing commentaries, especially in the two decades which followed the First World War, led to the Journal becoming known, pointedly by some, but affectionately by others, as "the yellow peril". The golden colour of its cover during this period may also have had something to do with this unofficial title.

Edward's legal background gave him a special interest in mining law and, more particularly in his earlier years, he wrote extensively on safety and health conditions in the industry. Under his editorship, the Journal's news service from special correspondents in all the principal mining fields became perhaps its most notable feature. Instinctively, he seemed to sense the right questions to put to his correspondents and in this way built up an ever-moving picture of local events set against their broader implications for the world industry. In his later years he acquired an international reputation as a minerals economist, particularly in tin, copper and the precious metals.

His editorship, being a highly personal one, reflected his own interests and specializations, and at times it appeared that insufficient attention was being given to the strictly technical aspects of the industry. The truth was that Baliol Scott was determined to sustain the essential qualities and functions of a weekly industrial paper so vigorously enunciated by its founder a century earlier. More than any other editor since the death of Henry English, he was a crusading knight, tilting sometimes at windmills, it is true, but always campaigning for a better, safer, and more productive industry.

Nevertheless, during the first few years of Baliol Scott's editorship, when he was busy shaping the Journal's 20th century policies, there were several influential British and American mining personalities in London who felt that too much space was being devoted to investment and general economic and political aspects of the industry, and too little to the practical problems of mining and engineering.

Among them was Herbert Hoover, who had by now become one of the world's leading mining engineering consultants, and for whom London was one of the principal bases of his expanding business. Others included several members of the IMM, which had now outgrown the cradle which the *Mining Journal* had provided for it and was becoming a flourishing organization of the industry's professionals.

Hoover took a leading part in getting the necessary money together to form a new company, Mining Publications Limited, for the establishment of a "real" mining magazine in the British capital. It was also Hoover's personal association with the Rickard family that led to the sponsors' invitation to T.A. Rickard and his business manager cousin, Edgar, to come to London from San Francisco to launch the project.

Under the Rickards' direction, the *Mining and Scientific Press* had made a remarkable recovery from the earthquake disaster of 1906 and was now well established as a highly enterprising and profitable undertaking. The Rickards felt that they could safely leave the paper to a caretaking staff—and, in any case, T.A.'s restless spirit could not resist the challenge of Hoover's invitation—and in June 1909 they moved with their families to London.

Before leaving San Francisco they arranged with their London correspondent to issue a sheet under the name *Mining and Scientific Press* in order to hold the title, but on arrival they found that others had registered a similar name and alternative likely titles to block them, so the new publication became simply *Mining Magazine*.

The first issue appeared on September 15, 1909. In size and general appearance, it looked American—almost a tabloid, compared with the "broadsheet" format to which both *Mining Journal* and *Mining World* still clung. Well printed on calendered paper, with plenty of illustrations, it also had a revolutionary style in its editorial utterances. Seven pages of "Review of Mining" were followed by 14 pages of "Editorial". Then came 12 pages of "Special Correspondence" from Johannesburg, Mexico, San Francisco, Denver, New York and Cornwall. Two pages of "Discussion" followed. J.H. Curle wrote on "West African Mines", H.F. Collins on "Sintering of Copper in Spain" and T. Lane Carter on "Metallurgy of the Rand". The issue also contained a technical description of the first flight across the Channel by Louis Bleriot, for which Rickard had obtained details from A.E. Berriman, a leading authority on aviation and then editor of *Flight*.

Editorial policy was expounded in a style which was as refreshing as an Atlantic breeze to those who were beginning to tire of the verbosity of the Victorian era. "The purpose of this periodical", wrote

Rickard, "is to be useful to those engaged in mining. To be useful a publication must be interesting, and to be interesting it must be truthful. By affording accurate and unprejudiced information on the various subjects pertaining to the technology and business of mining, we hope to win the support of the large number of earnest men engaged in one of the basic industries of civilization . . ." Then a jolt to his London contemporaries who never imposed any restrictions on the private activities of their employees: "Members of the staff of this magazine will not buy or sell mining shares, nor do they own any at this time. This decision is the result of experience elsewhere. Journalism has suffered from the temptations of share-gambling and is suffering from them today." And again, probably with *Mining World* in view: "It will be no part of our purpose to tell people what shares to buy and sell, although we expect to give a great deal of information likely to prove helpful to those who want to know the physical condition of particular mines or the status of the industry in particular districts. Mining papers of the better sort do not thrive on the support of the dabbler in the margins of the Stock Exchange, nor do they depend on the subscriptions of other persons of a nondescript character; the monetary success of the best type of technical publications depends on the fact that they are valuable media of publicity to those having machinery for sale. The stock-jobber, the punter, the shareholder, the other human units connected with the non-technical side of mining do not contribute to the success of a mining periodical as much as one manager who orders a winch from a firm advertising in that periodical."

In the light of the rather pious discourse about dabbling in mining shares, it was a little surprising to see in the same issue an anonymously-written article entitled "Investments and Speculations", introduced as "an appraisal of the present value of mining shares, contributed by a speculator unusually well qualified to form a just estimate".

There was great speculation on both sides of the Atlantic as to who this "speculator" might be, and surprise at the apparent anomaly of including such an article in a paper that had gone out of its way to emphasize that it would avoid such coverage. The *Canadian Mining Journal* wrote: "The *Mining Magazine* treads on dangerous ground in giving this class of article ... though published in all good faith, material such as this is extremely open to misuse. We agree heartily with the editorial diction of the *Mining Magazine* itself that it is best to leave stock market tipping to the financial periodicals."

It was not until some years later that the identity of the author was revealed. It was Hoover—the man who had inspired the launching of a magazine which would refrain from discussing matters of this sort and concentrate on the technical interests of professional engineers! The whole thing could have been cooked up by Rickard in order to start people talking about the new paper—and this they certainly did, with a considerable amount of "tut-tutting" within the hallowed walls of London's clubs and institutions.

Commercially, *Mining Magazine* was on its feet and earning a profit within a year, though Rickard had over-estimated the advertising potential and had failed to understand the subtle differences between advertising practice in America and in Britain. He always complained that British manufacturers of mining machinery did not appreciate the value of advertising to the same degree as the Americans. He deplored what he considered to be the "insidious" practice of British journals publishing the reports of annual general meetings of the big mining companies, more particularly the chairman's speech, in the body of the reading matter, and charged to the companies as advertising.

Years later, after he had retired from publishing, he elaborated on this aspect of his "London experience" which always seemed to distress him. "This sort of business is a sop to Cerebus", he declared. "It serves a sinister purpose because any editorial criticism of the management of a company is likely to be followed by a cancellation or withholding of the order for the advertising space required in reporting the next meeting of the company, or one of the same group. I used to attend the more important of these annual meetings myself and write editorials on them, describing the incidents and recording impressions as if I were telling someone at a distance what really happened. That was real journalism. I ignored the accounts given by the secretary of the company to the Press; I told the true story which often happened to be amusing as well as significant.

"As an example of the sort of thing that happened, I may mention the annual meeting of Consolidated Gold Fields of South Africa in 1913. Several attacks were made by shareholders against the board of directors, but they were poorly delivered. The Gold Fields people had been doing some shady business in share speculation. I said editorially: 'Lord Harris (the chairman) took the unction to his soul that he and his colleagues were acknowledged as fair and honorable in their corporate dealings in mines. We happen to know that the general opinion of the mining profession, particularly those acquainted with the performance of the Gold Fields in Rhodesia and Nigeria, is nothing like so kindly.' This was true, and it was said in the public interest. A few days after our issue of December 1913 had appeared I received a call from the secretary of Lena Goldfields, another of the companies of which Lord Harris was chairman. He upbraided me for giving offence to my 'friends' and at the same time cancelled an order for 3½ pages for a report of the Lena Goldfields meeting. I told him that by 'friends' he must mean 'patrons', that he entirely misunderstood our kind of journalism, which depended for its financial success not upon the patronage of promoters, but upon the fact that it was read by the mining engineers who placed the orders for the machinery that was offered for sale in our advertising pages; in other words, it was a medium of publicity for manufacturers, not a dependant of the *haute finance* of stock-jobbery."

Mining Magazine lost much advertising revenue in its early years because of Rickard's refusal to go along with this custom. Whether or not such practice can be held to be ethically acceptable is debatable, and such debate must also extend to any form of advertising that is allowed to influence editorial. But right or wrong, Rickard stuck to his principles and would have nothing to do with it. He also championed in his editorials the rights of shareholders, insisting that a director was not a privileged speculator with the earliest access to information that raised or depressed the shares of his company, but a trustee for all shareholders, who were entitled to any news from the mining properties without delay.

It is difficult precisely to assess what impact Rickard's magazine had on the two established mining papers in London, *Mining Journal* and *Mining World*. It could hardly have been a coincidence that within a few weeks of Rickard's arrival on the scene, *Mining Journal* knocked two inches off the width and five inches off the depth of its format to bring it more into line with what was becoming a standard size in the United States, Canada and elsewhere overseas. In doing so, it admitted to "some surprise to find, as a result of inquiries, that in the many mining offices filed copies of the Journal are not kept because ordinary shelves are not big enough to hold them". It had also taken the publishers well on 75 years to appreciate that mining engineers and other readers were frequently on the move and to carry around such a gigantic sheet, filled to capacity with such small type, was not of the greatest convenience.

Certainly, the very fact that there was competition, and that it was intelligently directed, was stimulating to the established London mining Press, and readers of all papers benefited by a wider spread of news and some more perceptive commentary. One of *Mining Magazine's* biggest triumphs was, of course, its publication in 1912 of the Hoovers' English translation of *De Re Metallica*, which brought it tremendous prestige at home and abroad. It also cemented a friendship between Hoover and Edgar Rickard who supervised its printing.

However, whilst *Mining Magazine* set new patterns in mining journalism and steadily increased its circulation, the *Mining and Scientific Press* was suffering acutely from Rickard's absence, both editorially and financially. When war broke out in 1914, Rickard, still an Englishman, wanted to remain in the country until it was all over, and he hoped to persuade Edgar to go back to San Francisco and set things right. But Edgar quickly became personally involved in Hoover's humanitarian work in Europe, first on behalf of American refugees, and then for the Belgians, and his cousin realized that he himself would have to return to the States if the *M and SP* was to be saved. He therefore exchanged his stock in the *Mining Magazine* company for Edgar's shares in the American paper, and in March, 1915 he sailed for San Francisco.

Without the Rickard touch, *Mining Magazine* soon slumped into mediocrity, a process which was, of course, accelerated by the restrictions imposed by the War. In retrospect, it may well be that its main fault was that it was some 50 years ahead of its time, both in its format and in its presentation of news and opinion in the punchy English, of which Rickard was a master.

Looking back on the magazine from San Francisco a few years after his departure, Rickard lamented that it was "going the way of all London mining and financial journalism in that it is becoming dependent more on the so-called company meeting advertising, which has the effect of silencing criticism and eliciting compliment, than upon straight advertising from manufacturers of mining machinery".

It is a pity that its first editor cannot be allowed a peep at the magazine today to see the fulfilment of his ambition in a well-produced publication supported by the dignified, but colourful, advertisements of the world's leading mining machinery manufacturers and many others who provide essential services to the industry.

CHAPTER ELEVEN

T. A. Rickard
– a profile

"... a pepper pot of a man, a burr under the saddle"

Thomas Arthur Rickard, mining engineer, journalist and author—one of the most prolific writers on mining affairs during the past 100 years—was born in 1864 at Pertulosa, close to the naval base of Spezia, in northern Italy.

For a man who was to travel the world extensively throughout his life, the circumstances of his birth were to demand frequent and sometimes tedious explanation to immigration officers of little comprehension. If he had been born in Italy, why, then, was he not Italian? Even when he applied for American citizenship in 1917, the judge lifted an eyebrow and mused: "Born in Italy, I see, Mr. Rickard?" "Yes, your Honour, but chiefly for one reason—my mother was there at the time. Your Honour will agree that there are two rules to which there is no exception: a man must be present when he is shaved, and be with his mother when he is born."

His brashness might have cost him the citizenship which he had deferred seeking until the United States had entered World War One on the side of the Allies. Fortunately, the judge, too, had a sense of humour and Rickard, Scottish-Irish on his mother's side and Cornish on his father's, was admitted to the citizenship of the country to which he had given so much of his talent, energy and initiative for more than 30 years.

His father's family was well known in a county which had contributed vastly to British mining and which, in the 19th century, not only became one of the most important centres of world copper and tin production, but sent its sons to the furthest corners of the earth, teaching others the skills and technology of mining. Thomas Rickard, his father, native of Porthtowan, near Redruth, was the eldest of five brothers—William Henry, Richard White, Reuben and Alfred were the others—who were all mining engineers and metallurgists. His grandfather, "Captain" (mine captain) James Rickard, was one of the first to go to California in the summer of 1850 on behalf of John Taylor and Sons, to appraise a group of gold mines on the Mariposa estate, then under option from General Frémont, "pathfinder" of America's West, and candidate for the Presidency in 1856. Captain Rickard took with him the component parts of the first stamp mill to be put to work in the Californian goldfields and can thus be accepted as a pioneer of American mining.

When T.A. was born—he was always known more by his initials than by either of his Christian names—his father was in charge of metallurgy at the silver-lead smelting works of the Societa di Pertulosa, then operated by a British company and later by the French Peñarroya group, which continues to have a controlling interest in the plant to this day. Two years after T.A.'s birth, his father went to Andeer in Switzerland for John Taylor and Sons and, after two years more, became manager for the Russian Copper Company, operating several groups of mines and smelters in the Ural Mountains and across the border in western Siberia.

The first language which the young Rickard spoke was German—largely because the nurse who accompanied the family to Russia was a German-speaking Swiss from Zurich. His first tutor was a Russian, and throughout his early childhood he thought in German and spoke in Russian better than in English. He accompanied his father on some of his inspections of the mines and acted as interpreter, not only for his father but for visitors from abroad, including Professor George W. Maynard, of New York, with whom he resumed acquaintance in Denver some 22 years later.

When he was 11 he was sent, with his sister and his brother Forbes, to school in England. The journey was an adventure for the children, taking three weeks from the Siberian border, up the Volga from Ufa to Nijni Novgorod and thence by train via St. Petersberg. In England, the two boys began their formal education at the Wesleyan College (later Queen's College) at Taunton in Somerset. T.A. quickly proved to be an above-average pupil; he found it easy to learn, and to win prizes; at 17, he had matriculated and gained a scholarship to Cambridge, and was about to go there, with a view to adopting one of the "liberal" professions, possibly the law, when he was diverted by the advice of his uncle Reuben, who arrived in England on a visit from Berkeley, California. Uncle Reuben urged him to follow the family calling, and T.A. had so much respect for his sagacity that he did as he suggested. And so did Forbes. T.A. enrolled at the Royal School of Mines in 1882, and a year later his brother joined him, the two boys travelling each day from Putney to South Kensington on a "double tricycle" until an accident put an end to this novel means of transport.

At the School of Mines, T.A. immediately found himself under the spell of one of the most remarkable men of the Victorian age, T.H. Huxley, biologist, philosopher and dean of the faculty. "Darwin's Bulldog" was at this time at his most controversial, energetically pursuing his campaign against orthodox beliefs. Though still a boy, Rickard learned quickly to appreciate that he was in touch with one of the great thinkers of the world and there is no doubt that Huxley had a profound influence on the moulding of his character and his approach to the problems which he encountered later in his career, both as a mining engineer and as an editor. It was Huxley, more than anyone else, who turned him into "a pepperpot of a man, a burr under the saddle".

Throughout the rest of his life Rickard would recall with pride that whilst at the RSM he served on the same committee as Huxley and was able to enjoy, if only in a small degree, a personal relationship with him. He was lucky to be chosen as students' representative on this committee, his appointment being partly due to the fact that although he had entered as a private student, he subsequently won a free scholarship and received financial assistance from the government, thereby ranging him alongside men who were being trained to be science teachers. Among these teachers-under-training was the poor boy from the draper's shop, H.G. Wells, who, according to Rickard, complained that the government subsidy was not enough to ensure him an adequate lunch and that in consequence he suffered from defective nutrition. As a group, the trainee-teachers were serious and hard-working, their future careers depending largely on their successful efforts as students. The private students, financed by their parents, took things more light-heartedly, and spent more time and money on sports, dances and other relaxations. Inevitably, there was some lack of comradeship between the two groups. Many years later, in a discussion in the *Mining Journal* on the status of the Royal School of Mines, Wells wrote that he remembered Rickard as "one of the Olympians who were unsociable to my group".

If Wells was disgruntled with his experience at the School, Rickard certainly was not. He enjoyed every moment, never failed any examination and obtained a first-class degree. In those days the lectures dealt more with principles, and less with the practical application, of mining; it was more a college of science than a school of mines. Rickard was quick to recognize this. Being the son of a mining engineer and having heard mining discussed at home so much, he skipped the mining course and took the one in metallurgy, having made up his mind that practical mining was not a thing that could be taught in lectures. As a result, he missed the benefit of Sir Warrington Smyth's great teaching which did much to balance the sturdy merit of the fine old mine captains of the day with a modicum of theoretical and applied science. Instead, he filled in his time by going to Westminster to listen to the debates in the House of Commons. He heard Gladstone in one of his impassioned moments, Arthur Balfour, Lord Randolph Churchill and other distinguished men of Victorian politics. But the great benefit that he derived from his three years in London remained the stimulus imparted by mental contact with Huxley. "His lectures", wrote Rickard many years later, "were windows to the infinite."

*

Rickard graduated from the Royal School of Mines with the class of 1885, and in June of that year sailed for the United States to join his Uncle Alfred, youngest of his father's brothers, in Colorado. Alfred,

133

a kindly man who to the young graduate seemed more like an elder brother than an uncle, was manager of several British companies operating near Idaho Springs and Central City. Alfred hired his nephew as assayer at $100 a month, and Rickard received his first pay cheque on his 21st birthday. The following year, arrangements were made for him to spend some time at Leadville to learn surveying and, on his return to Central City, he was appointed surveyor as well as assayer—and his pay was doubled.

Then suddenly, and tragically, Rickard found himself in temporary charge of all the properties under his uncle's management. His aunt died in childbirth and Alfred returned to England with the rest of the family of four children. This was the first test of the young man's qualities of management and responsibility. He took it in his stride, but not without some anxious moments. One of the mines which came under his supervision was the California, then at 2,000 feet or more, the deepest mine in Colorado. Hoisting was done with a hemp rope and Rickard would catch his breath as it slipped on the drum, giving the bucket a drop which he feared might send it to the bottom.

The environment in which Rickard served his mining apprenticeship was vastly different from the Victorian gentility in which he had spent his schooldays in England. Here was the Wild West at its wildest and most crude. Leadville was a particularly lawless town. Rickard did not really have the stomach for it. He could find no romance in the violence and uncouth behaviour of so many of the miners and pro-spectors and their vast army of camp-followers. Almost every day there was a shooting fray in one of the townships, but it did not interest him. He never carried a gun, like most of his companions, and avoided arguments which could give rise to physical conflict. Despite a mental abrasiveness which had been stimulated by his contact with Huxley, he was essentially a delicately-nurtured youth, believing sincerely that the quest for riches was incidental to the scientific search for ore and the application of technical ingenuity to turn it into something of use.

During his uncle's absence in England Rickard had a number of visitors, among them Philip Argall, recently appointed manager of the La Plata company in Leadville where Rickard had learned his surveying. Argall, then only 33, had come from a similar background to Rickard's—Cornish on his father's side, Irish on his mother's. At 22, he had married into another well-known Cornish family. His wife was a daughter of Captain George Oates, who came from a long line of copper and tin miners, and sister of Captain Lawrence Oates, of the Inniskilling Dragoons, who walked out of his tent to his death in an Antarctic blizzard to relieve the burden on his companions during Scott's retreat from the South Pole in 1912.

Rickard's father had in fact been responsible for Argall's appointment as manager of the Duchy Peru mine, near Newquay in Cornwall, some eight years previously, and it was here that T.A., then a boy of 14, had met him for the first time and bombarded him with questions about the life of a mining engineer. Argall later went to New Zealand, again at the recommendation of Rickard's father. Now Argall was to return a family compliment. Impressed with young Rickard's work in Central City, he asked him if he would like to be manager of a gold mine which he had recently been examining in California. Rickard jumped at the chance (he had spent some time the previous winter with his Uncle Reuben at Berkeley and had returned with a delightful impression of California) and his first managerial appointment followed—to the Union Mine, a few miles south of the road to Angel's Camp where Mark Twain's famous story of "The Jumping Frog of Calaveras County" had its genesis 22 years earlier. Rickard took charge of the Union Mine on his 23rd birthday, acquiring at the same time a house, hot and cold water, a Chinese servant, a horse and buggy, and $300 a month. But the mine itself was, in Rickard's words, "a fizzle". It had been sold to an English company solely on the recommendation of its promoter who, Rickard discovered, used to paint cuspidors and coal-scuttles for the Southern Pacific Railway and, as part of the deal, had remained manager of the mine at a handsome salary. Argall had exposed the whole set-up and the promoter was fired. Rickard was able to improve production temporarily and received complimentary letters from the directors, but when the returns dwindled he was upbraided for his failure and, early in 1889, a mining engineer, an Australian, was sent out from England to investigate conditions and to examine his management. The visiting inspector not only exonerated Rickard from all blame, but asked him to go with him to Australia. The offer was accepted and in October the same year they went together to London, and from there to Australia.

Rickard's job in Australia was to examine mines which his sponsor was promoting; regretfully he had to report unfavourably on most of them, but they remained good friends. He had saved enough money from his salary in California to be able to travel, and this he did energetically. In the course of two years he examined more than 80 mines, all the way from Broken Hill to Mount Morgan.

Although he received precious little in the way of fees, Rickard obtained a good working knowledge of the principal ore deposits of Australia and of the prevailing mining and metallurgical processes. More important to his future career, he began to write more frequently about what he saw in his travels. Rickard had always loved writing. At school he had been a regular contributor to the magazine and had won prizes for verse. He began writing technical articles when he was at the Royal School of Mines and, whilst in Colorado, he had an article on the ore deposits at Redcliff, near Leadville, published in the London *Mining Journal*. In Australia, he wrote several articles for the Melbourne *Evening Standard* and also made his first submissions to the Transactions of the American Institute of Mining Engineers. An article on the Bendigo saddle-reefs was one of his earlier contributions to economic geology, and for many years, if he was known at all, it was as the author of that paper.

It was a particular pleasure for Rickard to write for the AIME. Dr. Raymond had by this time left the *Engineering and Mining Journal* to become secretary of the Institute and editor of its publications, and Rickard benefited greatly from his help and guidance. He was his principal tutor in technical journalism. Raymond would send him letters, 10 or 12 pages long, explaining corrections to his manuscripts and making suggestions for improvement. From him, Rickard learned the art of presenting technical material in an attractive and readable way without losing the dignity of the subject; he had a keen feeling for the English language—despite his early tuition in German and Russian—and was never slipshod or slovenly in his writing.

Towards the end of 1891, whilst in Otago, New Zealand, Rickard received a cable from his father asking him to go to France to take charge of a project, of which he was managing director, to develop two groups of cobalt-silver and lead mines in the beautiful south-eastern part of the country. Rickard lived in regal luxury in a house which had belonged to the Dauphine two centuries earlier but, as he was later to remark, "the Lord rarely puts mines in such lovely surroundings". He soon lost confidence in the enterprise and returned to the United States to help examine projects on behalf of H.H. Warner, who was engaged in various mining operations in the Western States. On Rickard's advice, Warner bought the Hillside mine, near Prescott, Arizona. "But then," declared Rickard in a later interview, "to my horror, he started a share campaign characterized by practices of a questionable kind. The mine itself was a good one, but he capitalized it at several times its value and then proceeded to gut it."

Rickard resigned and opened an office on his own account at Denver as a consulting engineer. The next 10 years were busy and rewarding. For six of those years he was State Geologist of Colorado, and in 1897 he was engaged by the Venture Corporation, of London, to examine mines in Western Australia, then in the grip of a gold boom centred on the rich deposits found at Kalgoorlie and Coolgardie. Nevertheless, a great many "wildcats" had been foisted on the public, particularly in London, on the strength of the genuine mines, and the organization for whom he was acting had acquired much property that was no good.

On this second visit to Australia, which lasted about a year, Rickard had his first meeting with Herbert Hoover, who was then working for Bewick, Moreing and Company. Hoover was then 23 and Rickard 33. "We became", said Rickard, "pleasantly acquainted, and we agreed to exchange reports on wildcats; that is to say, we let each other know of the prospects that we turned down so as to avoid useless examinations."

Rickard returned to Denver just when the United States declared war against Spain in April 1898. In the same year he married his cousin Marguerite, daughter of Alfred Rickard, who had given him his start to his mining career 13 years earlier. In the same year, too, Rickard engaged a new assistant, fresh from college, in his Denver consultancy. His name was Alfred Chester Beatty, later to become one of the great mining engineers, "father" of the Northern Rhodesian (now Zambian) Copperbelt, a British knight and philanthropist. And when Chester Beatty embarked on matrimony he, too, turned to the Rickard family, choosing as his bride Marguerite's sister Grace Madeline, whom he married in 1900.

Rickard's work as a consultant made it necessary for him to travel a great deal. For 12 years he averaged 35,000 miles a year—by train and ship, it must be remembered. During the first year after his marriage he was home only once for a whole week. He found the call for travelling on such a scale to be increasingly unpleasant for, as a newly married man, he wished to be at home far more frequently. Moreover, whilst his technical work was absorbing and satisfying, he began to find that as he was called on to assess bigger and bigger mining projects he was increasingly obliged to take part in the negotiations associated with their purchase. For this he had little taste, but involvement with directors, promoters and stockbrokers was almost inevitable: he could not always pick his clients and he had to face realities.

It was these two considerations which influenced him most when he received an opportunity to turn to full-time journalism while still remaining within the parameters of the mining profession. Shortly before his death in 1901, Rothwell had asked him to join him as associate editor of *Engineering and Mining Journal*. Raymond, who had shared the editorship with Rothwell for some 15 years, advised him against accepting, partly because he felt that Rickard would find it difficult to work with Rothwell, and partly because the salary then offered was very small compared with his earnings as a mining consultant.

In the chaotic period which followed Rothwell's death, control of *E/MJ* passed first to James H. McGraw and then to William J. Johnston. Johnston had previously offered Rickard the editorship of a paper called *Mining and Metallurgical Journal* in Los Angeles, but Rickard had told him that the only paper for which he would consider leaving his consultancy was *E/MJ*. Johnston remembered this when he bought the Journal from McGraw and consolidated with it the Los

Angeles paper: he wrote to Rickard, making him an offer of the editorship, which was promptly accepted.

*

Rickard began his new career in New York on January 1, 1903—and soon discovered that editing a weekly publication of the size and reputation of *E/MJ* was a vastly different proposition from being a technical writer accustomed to preparing at leisure occasional contributions on subjects of his own selection. The world into which he had plunged was a much larger one than that encompassed by the activities of a mining engineer and he had to interest himself in all sorts of new problems and controversies. He realized in his first week that, despite Raymond's diligent coaching, he was, both technically and professionally, untrained for his post. He had to learn quickly—and he did so, largely because of the help he received from Frederick Hobart, who had occupied the chair for a few months immediately after Rothwell's death, and from Walter Ingalls, who was a regular contributor and became Rickard's successor as editor when he resigned. Hobart was already an old hand at the Journal. He was not an engineer and he knew little about mining, but he understood the technique of editing and publishing and was one of the chief props of the paper until he retired at the age of 75. He taught Rickard all he knew about the complexities of class journalism.

Unfortunately, however, there were other complications—financial complications and problems over ownership and control. Johnston, like many other New York publishers of his day, had an obsession to own a string of papers and he attempted to do so, indiscriminately and disastrously. Crisis followed crisis in the *E/MJ* board room and control of the Journal changed twice more over the editor's head within the next 18 months. Rickard had been able to work with the first two owners, but not with the third—John W. Hill, who acquired control in 1904 and who was to merge with McGraw some years later to form the McGraw-Hill publishing company. Rickard found Hill, formerly a locomotive driver, "a man of integrity and ability, but rough and extremely domineering", and he soon gave notice of his intention to resign.

There seems little doubt that had he so wished, Rickard could have himself obtained control of *E/MJ* during this unsettled period. On his appointment as editor, he had invested a fair sum of money in the paper and during Johnston's financial indiscretions he had been responsible for setting up a committee of investigation. Johnston was allowed to remain president of the company, but the actual control of affairs passed largely into the hands of Rickard and J.T. Morris, the business manager. This was a wretched time for Rickard; struggling to get on top of his editorial job, he never knew from day to day what new difficulty he might have to face because of Johnston's vagaries. Towards the end of 1903, the committee, which consisted predominantly of a number of eminent mining engineers, urged Rickard to give his support to a plan designed to obtain control of the paper for the profession. But he wanted nothing to do with the publishing end of the business. All he desired was to learn the craft of an editor, and to be an independent editor. He felt strongly that if the paper were to be controlled by a directorate of mining engineers it would be a coterie in New York—a coterie that either individually or collectively, would expect to direct policy, not for any wrong purpose, but still in a measure for their own interest, or for that of their friends.

Rickard had also already acquired a distaste for what he called "institutional journalism", the running of any subsidized periodical without the discipline which comes from explicit responsibility and the need to be financially viable. In later years he was to attack most vigorously the magazine of AIME and those of other professional institutions, including the Canadian Institute of Mining and Metallurgy. It seemed to him that such bodies were no more justified in engaging in publishing activities than in running an ore-testing works or an assay office; producing magazines was foreign to their purpose and could be better left to those outside who had devoted themselves to this kind of work. He felt there was some excuse for the *Mining Congress Journal* because the American Mining Congress was, in essence, a political lobby in Washington, working on behalf of the mining industry; even so, he thought that the leading technical papers in the United States could combine to give the publicity that the Congress required.

Such views were already beginning to turn him into a controversial figure in his early days in professional journalism. He believed firmly that the urge to preach and to criticize was the essential of successful editorship. Already, too, at *E/MJ* he was following a self-imposed rule not to own any mining stock, not to be a director of any mining company, nor to participate in any mining promotion. He had made up his mind that an editor, in order to retain his independence and self-respect, must be in an unimpeachable position. To that policy he remained true throughout his career, and it proved a major factor in his success. He imposed the same rules on his staff—and the only man to break them paid with his life: one of his news editors became interested in a prospect owned by an ill-tempered old miner who was working as a night-watchman on a railway; he went to see him one morning, a quarrel broke out, and the editor was shot dead.

Before he left *E/MJ* in 1905 Rickard had already begun negotiations to buy the *Mining and Scientific Press* in San Francisco. Still chary of getting too heavily involved in the publishing business, he always asserted that he acquired the paper solely to enable him to become an independent editor. When he assumed control he appointed his cousin Edgar—a son of Uncle Reuben—as business manager and this proved a most successful arrangement. T.A. and his wife set up home, pleasantly situated at Berkeley (another cousin, Tom, was mayor of the town!), near the University of California, on the eastern shore of the Bay, and settled down to face their new challenge. It came sooner than they expected. The year was 1906, the year of the San Francisco earthquake disaster.

Shortly before 5 o'clock in the morning of April 18 the Rickards were rudely awakened as their home rocked as if at sea; the chimney collapsed and bricks rattled on the roof and verandah. Looking outside, Rickard could see the trees swaying and the dust rising from the stricken ground, "but", as he wrote later, "it was too early for breakfast so I went to sleep again". Worse was to come. After breakfast, when about to start for the city, now visibly on fire, there came two shocks in close succession. To Rickard, they felt and sounded like blasting in a mine a hundred feet deep. While discussing with his wife whether or not to attempt to cross the Bay, there was a phone call from Edgar to say that he was at the Mole; the ferry boats were not running but he would make his way as soon as he could.

"It was several hours before he reached our house," recorded Rickard, "and during the interval I went to my little study and wrote an article on earthquakes so as to have something pertinent to the occasion ready for publication. As soon as we met I said that all we needed was an office and printing facilities, for of news there was enough. Fortunately, we had a complete list of our subscribers and the dates on which their subscriptions terminated. This was recorded on a stencilled paper ribbon that we had been in the habit of taking home, one copy to my house and one copy to Edgar's. We decided to establish ourselves in Berkeley, for already it was apparent that our business quarters in San Francisco were in flames."

Edgar engaged a suite of rooms on the top floor of an undamaged bank, arranged for printing with a local paper and engaged a number of refugee girls to address wrappers. In this way the *M and SP* was able to send a single page of greeting to its subscribers before the fire had stopped spreading in San Francisco: " . . . our plant has been demolished, but this paper is built on nothing so ephemeral as paper, and on nothing as cheap as machinery; it is based on the support of many thousand readers and subscribers who are never less likely to withdraw their support than at a time of misfortune . . . there is another something that is not destructible by physical misfortune or financial adversity, and that is the spirit that gives life to the printed word."

Most of the people in Berkeley, which escaped serious damage, realized that the disaster had been due to a slip in the rocky crust of the earth, but at first they had no idea of the magnitude of the consequences. During the morning after the shock all sorts of rumours were heard: that Chicago was in the Lake, that Denver had been destroyed, and that a tidal wave had swept over Los Angeles. The telegraph wires were down and San Francisco was virtually isolated. It was lovely, sunny weather; nature seemed to mock the distressed inhabitants of the stricken city; the clouds of smoke rose nearly 5,000 feet and threw strange shadows over the waters of the Bay. More than a thousand people lost their lives and the financial loss was assessed at $500 million. But in some ways the earthquake episode was stimulating because the ill effects were soon forgotten and, like San Francisco itself rising from its ashes, the business of the *Mining and Scientific Press* was reborn in the ruins of its plant.

Tragically, however, the very act of reconstruction of the city disclosed a scandal of municipal corruption. Rickard joined the committee of the reform movement and spoke on the public platform. Among his special targets were the newspaper magnates, William R. Hearst and Michael De Young. Hearst controlled the San Francisco *Examiner* and De Young the *Chronicle*, indicted by Rickard as being "newspapers of the vilest type, owned by men without character and edited by men without principle". He accused them of hindering the punishment of graft and the city's orderly reconstruction. He threw his own paper into the battle and—much to his delight—was placed on the *index expurgatorius* of both the *Examiner* and the *Chronicle*. Before leaving California a few years later, Rickard received from the Citizens League of Justice an American flag and a citation that, "though a citizen of another land, you have stood with bold invincibility at all times for the right and honour of this community . . . your virile and able pen spoke week by week in clear and certain note . . . the whole force of your strong personality and the power and influence of your recognized place in our community were given unstintingly to serve the moral and civic life of the city, waging desperate battle for its honour, with the issue most uncertain."

Heady words! But these were heady times. Meanwhile, the main function of the *M and SP*, to serve the interests of the Californian mining man, continued to flourish. When Rickard took control of the paper he placed a motto on the title page, "Science has no enemy save the ignorant", reflecting—a little puckishly, perhaps—that such a motto would have pleased Huxley.

With Edgar, he continued to work enthusiastically; circulation and advertising grew steadily, revenue and profits increased. Rickard was always at pains to explain to those unfamiliar with the economics of publishing, that subscriptions from readers did not meet more than a fraction of the mechanical costs of publication, the main source of revenue being the advertisements. He claimed that in effect this contributed to the independence of the editor because, so long as he published matter that was read by those whom the advertisers wanted to reach, it did not matter whether the editor was popular or otherwise. "As I had no dealings with advertisers", he said, "I had a free hand and endeavoured to print what would engage the active interest of the mining public, more particularly the mining profession, because, of course, we had nothing to do with share-dealing, except sometimes to expose nefarious performances."

Some idea of the high reputation which the *M and SP* enjoyed at this time can be gained by the names of the special correspondents whom Rickard had gathered together to help him provide his readers with the best thinking and the best writing available on mining affairs. The list included Herbert Hoover, in London; Philip Argall, whose stature both as a mining engineer and technical writer had grown enormously since his meeting with Rickard in Colorado 20 years earlier; Walter Harvey Weed, who was later to acquire and edit Horace Stevens's famous *Copper Handbook* when the latter died in 1912; James Ralph Finlay, secretary of the Mining and Metallurgical Society of America, and lecturer on mining economics at Harvard University; Horace V. Winchell; J. Parke Channing; James F. Kemp; and others of similar repute.

One of the results of this assembly of talent, expertly handled by Rickard and translated into attractive journalism, was that the readership of *M and SP* became more and more international. It was not surprising, therefore, that Hoover invited him to London in 1909 to take charge of the new *Mining Magazine*.

Rickard summed up his six-years stay in London (described in the previous chapter) as "good fun ... an enriching experience, particularly of men and mines in a world-wide way, for the mining finance of London has a geographical amplitude like that of no other city in the world." Before he returned to the United States he received a presentation from the Institution of Mining and Metallurgy, from whose council he had resigned when he started *Mining Magazine* in order to be in a position freely to criticize its policy if he felt such criticism to be warranted. "I suppose", said his old friend Fred Hamilton in making the presentation, "there is hardly a man in this room who, at one time or another, has not had occasion to differ with him on some subject, and when we differ with Mr. Rickard we know it. But I am quite sure that there is no man here who does not recognize that no word he has ever written has rankled or left a poisoned wound, that no word he has ever uttered has been actuated by any personal or unworthy motive, or by any desire than to present the truth as he understood it."

It was a fitting tribute to Rickard's work in London, one of the last he was to receive as an Englishman, for very soon he was to become an American citizen.

Back in San Francisco, he worked tirelessly for the restoration of the *M and SP* to its former high standard and financial viability. A month or two after his return a telegram came from John A. Hill of *E/MJ* offering to buy the paper for $25,000. Rickard replied with, "The compliments of the season". Within two years circulation had doubled and the paper was back in its stride. But by 1922 he had had enough of publishing and, ironically perhaps, he sold out to McGraw-Hill (for 10 times the amount previously offered by Hill) who promptly merged it with *E/MJ*.

In 1926, Rickard was offered the secretaryship of the AIME, the directors declaring that they wished to give the post the prestige which it had enjoyed during Dr. Raymond's distinguished tenure of office. But a clause in his agreement with McGraw-Hill that he should not engage in technical journalism in the United States, Canada or Mexico, other than as a contributor to *E/MJ*, prevented him from accepting. The clause seemed unreasonably restrictive for a man from whom words flowed so freely and who believed passionately that the maintenance of a strong technical press was an essential to the well-being of the profession as a whole. "As literature is the criticism of life," said Rickard, "so technical journalism is a criticism of the life of the technician ... it has done much to stimulate the development of character in the profession and to render the profession articulate to the public."

And so the great tide of articles, treatises, lectures and contributions to the "transactions" of the professional institutions, began to ebb. But the waves still crumbled on the shore, bringing longer and even more carefully prepared non-technical works. Rickard had already written several volumes on his travels throughout the world, among them *Through the Yukon and Alaska* and *Journeys of Observation*, the latter being the first book published in San Francisco after the 1906 earthquake, and based on his travels in Mexico and on horseback through Colorado. Now was to come his *History of American Mining* and his *magnum opus, Man and Metals* which McGraw-Hill published in two volumes in 1932. The latter was prompted by his reading of H.G. Wells's *Outline of History* because he felt that his former fellow-student had omitted to pay proper regard to the part that mining had played in the development of civilization. It was, for Rickard, a pleasing thought that Huxley, under whose influence both of them were placed half a century earlier, would have commended the highly educative value of Wells's work in *The Science of Life* and would have rejoiced to see it produced in collaboration with his own grandson, Professor Julian Huxley.

Disappointingly, Rickard had to decline an invitation from the children of Dr. James Douglas to write a biography of their father, one of the most eminent mining engineers and metallurgists of the age, and whom Rickard described as "the James Bryce of mining—kind, sagacious, public-spirited, an able man and a good citizen". Rickard had first met Douglas by chance, in the Pullman car of a railway train as it stood at the station at Tucson, Arizona, early on a spring morning in 1895 when the air was fresh and fragrant. He introduced himself: "My name is Rickard." "Ah, now, which one are you?" "T.A., son of Thomas." "Oh, yes, I know your father." What a tight little world it was in those days! Thirty years after that first meeting, Dr. Douglas's son Jim invited Rickard to accompany him on a journey through southern and central Africa, in the course of which they visited the area which was to become the Northern Rhodesian Copperbelt, the biggest concentrated copper-producing area of the world, developed on the initiative of Alfred Chester Beatty, who had started his career as Rickard's assistant in Denver.

Rickard always told his friends that if he were to be remembered at all after his death he hoped it would be as one who had helped to improve the writing of those in the engineering profession and to give new editors the grounding that he himself had lacked. As early as 1908, when he was still learning the basics of class-journalism, he published a *Guide to Technical Writing*, which went to several editions and finally, in 1932, was revised and selected as one of five books given, under the terms of the Seeley W. Mudd Memorial Fund, to 1,200 junior members of the AIME to help them in their studies. The book can still be found in some editorial offices and there are a few journalists of the old school who say that their copies continue to be well thumbed.

Unfortunately, the book is no longer "required reading" for those aspiring mining editors with no formal training in journalism but, comparing the standard of writing in some present-day publishing houses with that of 70 years ago, it may be appropriate to suggest that the time has come for a new edition. If it were done it would be in keeping with the spirit of a man who recognized two professions—mining and journalism—and served them both equally well.

CHAPTER TWELVE

On the Level in Canada

One day in March 1915, a few formes of hand-set type, put together in a basement fronting the muddy square of the mining town of Cobalt, in northern Ontario, were loaded carefully into a street-car to begin a 10-mile journey to Haileybury. The track was reputed to be the "jiggliest" in Canada and every bump and sway of the car was agony for those who accompanied the precious load. At the end of the ride the formes were carried into the printing works of a weekly paper called *The Speaker* and made ready for the press. A few hours later, a new paper was born, a paper that was, and still is, unique in mining journalism.

Looking back across 60 years which have seen events that have not only wrecked the best-planned business ventures, but have sometimes threatened the existence of civilization itself, the middle of a world war may seem to have been a strange time to start a mining paper. But the war had not then hardened into the holocaust that it was soon to become, and Cobalt was busy enough, and prosperous enough with its thriving silver mines, to be relatively unaffected by what was going on in the far-off battlefields of Europe. Moreover, the war would not last: the end was just ahead, so people believed. In these days of hope and fear, the *Northern Miner* made its first hesitant steps in an infancy more than full of the ills that usually befall childhood.

The founding fathers were Ben Hughes, who had contributed a mining column for the *Cobalt Daily Nugget*, and was probably one of the best mining writers of his day; and Ernest Hand, a printer of rare skill, who ran a small weekly paper in competition with the *Nugget*. At this time the prevailing philosophy applied to mining publications in Canada was that only good news should be publicized: "If you can't boost, don't knock" was the advice pressed upon aspiring mining reporters by those whose interests lay with the large companies. But Hughes and Hand felt otherwise: they were convinced that the public would welcome a paper that would tell the whole story, both good and bad. They also wanted to produce a professional paper that non-professionals would understand and buy, and, with these objectives in mind, they adopted a level as their emblem and "On the Level" as their motto, a reflection of editorial policy that has continued to this day.

In terms of tangible assets, the partners were not very well endowed. Neither of the men could bring much in goods or money to the business. Ernie had a few founts of type and some small items of printing equipment; Ben had a keen sense of news value and a facile pen. At first, they could not afford a press of their own—hence the weekly street-car ride to borrow one. To raise cash, they passed the hat round among mine managers, bankers and merchants, and the $50 and $100, and sometimes larger shareholdings of these men, brought to the struggling venture its first capital. When eventually there was enough money to buy and install a press there was no room left in the basement for normal access and the staff went to work through a trap-door in the ceiling.

The paper—produced in newspaper format from its inception—was just over a year old when Hughes, a patriotic Englishman, felt the urge to enlist. He sold his half-interest for a modest sum to Richard Pearce, who had been refused by the army for medical reasons. Richard at that time was acting as correspondent for various papers. He had been a reporter for the old Toronto *World* and during the course of an election campaign had written some uncomplimentary articles about one of the local politicians, who was demanding retribution from the paper and threatening a lawsuit. Richard was told by his editor to "get the hell out of town and stay hidden for a couple of months". He went to Cobalt, and it was a quarter of a century before he moved back to Toronto with the *Northern Miner*.

As the war continued, so the business problems became more serious, but somehow the paper came out on time every week. Wages for the staff were small—but they were always paid. Accounts were collected on a catch-as-catch-can basis; there was no time for professional accountancy. Richard's brother, Norman, who had also worked on the *World*, sensing the future of the mining industry, joined the organization in 1920 after his war service. One of his first tasks was to open a set of books during which he discovered that one advertiser had not been billed for three years. Timidly he sent an invoice which produced an immediate cheque for $300; this was a high day in the fortunes of the paper and the name of the advertiser, Hull Iron and Steel, was long emblazoned as the saviour of a trying situation.

The following year, another brother, John R. (Jack), joined the company and took over the business worries, leaving his two grateful brothers to indulge their passion for newsgathering. It is interesting to record that in the fullness of time each of the three brothers had one son, and during one period all six Pearces were in the business.

But of far greater significance is the fact that Norman and Richard Pearce, who shared the editorship for a quarter of a century, had both been trained in the hard, competitive school of the daily newspaper, and they stamped the pages of the *Northern Miner* with a journalistic expertise that had been rarely seen in any other mining paper up to this time—and which is still unusual today. Norman was also a fair hand with a camera and had been the first to take aerial pictures over Canada, from the first flying boat to operate in the country.

To a mining readership which had been used to the verbosity of the enthusiastic amateur, or to the conservatism of some mining engineers who had assumed the role of "editor" without training or experience, the *Northern Miner* at first provided something of a shock. Canadians were slow to accept the policy of full disclosure. As some people, even today, blame the media for bad news, so the *Miner's* readers blamed the paper for any upsetting disclosures. Copies were sometimes burned in public and the editors hanged in effigy. To the Pearces this was no discouragement; more was it a sign of success; their paper was having an impact, even if some of its readers did not always like what they wrote.

Already, before the First World War ended—and before Norman

Pearce had officially joined the paper—the bold treatment of a situation which the brothers uncovered resulted in great benefit to the Canadian mining industry. The story concerned German involvement in the payment of royalties for the flotation process, which had become a key factor in the economic treatment of ores at many mining properties all over the world. In simple terms, the process consists of reducing milled ore to a pulp, to which is added liquid and suitable chemicals; the mixture is agitated and air is blown in to produce bubbles, which carry fine particles of minerals to the surface, where they can be skimmed off and the waste "sink" discarded as tailings. Tradition has it that the discovery of this process was the result of the sharp eyes of a lead miner's wife who, when washing her husband's overalls, was intrigued by the fact that particles of glena, a lead sulphide mineral, always stuck to the soap bubbles in the tub and sparkled on the surface. There is some doubt about the truth of this story, which has been claimed by several mining countries, but a company was formed to register the patent of minerals separation by flotation and to protect the rights of the process, which soon became in world-wide use. It has been said that the flotation process made possible the development of the great porphyry copper deposits of the United States and South America in the early years of the present century—it certainly made them profitable—and the process was also a factor in the decision to go ahead and open up the Northern Rhodesian Copperbelt in the 1920s. In Canada, flotation had become important in handling the Sullivan ores in British Columbia and making the Consolidated Mining and Smelting Company a big power; the nickel mines needed it, and it had become essential to operations in the Cobalt camp.

To install the process was relatively inexpensive but, not unnaturally considering its origins, troubles arose concerning its use and the royalty payments demanded by those who claimed the rights. The basic patents were held by a British company, but Canadian royalty was payable in New York to a company known as Minerals Separation North American Corporation. Canadian companies had long considered that the royalties demanded of them were excessive. Whilst on leave in London from military service in France, Norman Pearce decided to look into company records, from which he discovered a relationship between the patent-holders and a German metals group. Further investigations by the *Northern Miner* revealed that Minerals Separation North America was controlled by another New York company called Beer, Sondheimer, blacklisted by the British Government under its Trading with the Enemy Act. The *Miner* at once began a vigorous campaign for the immediate annulment of the royalties, exposing Beer, Sondheimer as one of a group of interests dominating the German metal market, "representatives of the Kaiser in the metals world . . . enemies of the Empire".

Naturally, those Canadian companies from whom heavy royalties were being demanded and who were being threatened with litigation by Beer, Sondheimer, backed the *Miner's* campaign. Eventually, with the help of the Americans, soon to join the Allies with their immense resources, the enemy's grip on flotation was broken and the royalties were substantially reduced, not only for the Cobalt silver mines, but for base metal mines throughout the whole of North America.

The *Northern Miner* was always very conscious of the importance of Canada's mining industry to the war effort. Gold was urgently needed to pay for war purchases in the United States, which demanded "cash on the barrel", and silver was wanted by the British for settlement in the Orient—and to pay the Chinese labour battalions which worked behind the Allied lines. In the mining of silver, cobalt was a valuable by-product and contributed greatly to the munitions industry, especially when combined with tungsten and nickel to produce steelite. All base metals were in demand: prospectors and developers were relieved of war service to find new resources, and Cobalt, centre of mining activity in the north, took on a new usefulness and activity.

All this generated stories for the *Miner*—including many that the paper could not tell. Among the latter was the discovery that considerable quantities of nickel were finding their way to Germany through American ports. Canadian producers were helpless to prevent this. But the British Navy took action: they broke the German code of wireless messages and quietly intercepted many ships, loaded with nickel, copper and other metals, and diverted them to British ports. The Germans countered by running all their signals at top speed to make them unintelligible, but the British cryptographers found a means of unravelling them and, once again, metal which the Germans had paid for, was finding its way to British war factories. The *Miner* had the facts but could not print the story.

Norman Pearce, whose hunch had produced the flotation royalties story, had a distinguished military career in France; he served as a lieutenant in the 6th Brigade, Trench Mortars, and won the Military Cross. In the Second World War he played a big part in forming the Second Battalion of the Irish Regiment, which he subsequently commanded. But it is as a writer—perhaps even more than as an editor—that he will be remembered most. News stories and editorials flowed from his pen with a facility which astonished many of his contemporaries, among them some veteran newspapermen. He had the knack of turning a Churchillian phrase, without appearing pompous and, like the great war leader, he could marshal his facts and present his arguments in telling fashion.

A keen student of economics and monetary systems, he was one of the very few to recognize the true significance of events in 1931 when Britain went off the "gold standard". Through the columns of the *Miner*, he attempted to stem the panic that gripped sections of the Canadian financial community and the public. The wisdom of his advice was eventually accepted and the Canadian mining industry boomed as it never had before, while gold mining proved to be a bulwark of strength in a country struggling to overcome the effects of the years of depression.

The advent of new mining fields across the length and breadth of Canada quickly led to the *Northern Miner* becoming a truly national paper with a circulation which was eventually to challenge the best of its contemporaries in North America. Moves to larger premises in Cobalt were followed by a decision in 1929 to headquarter the ever-growing enterprise in Toronto, then rapidly becoming Canada's most important mining centre, its nearness both to Cobalt and the Ontario gold mines placing it in an ideal position to assist in opening up new mines across the country. For most of those on the *Miner* it was a sad parting, as Cobalt was a friendly and exciting place in which to live. But for the Pearces, the move to Toronto was coming home, for it was here that they were born and obtained something richer than the rudiments of the newspaper profession.

Working out of Toronto, the brothers extended their practice of making personal visits to mines and new prospects wherever they were located. Few other mining papers have so consistently gathered their news at the source and amassed such a knowledge of individual mines and mining conditions at first hand and, perhaps most important of all, made so many personal friendships with men working in remote areas, many of which were to become important place-names on the map.

The *Miner's* practice of relying on their "staffers" and not normally employing correspondents in the mining camps represents a fundamental difference of policy compared with most other mining papers. Again, this policy reflects the daily newspaper instincts of its founder-editors, more particularly, perhaps, of Norman who always argued that mining demanded the best available system of reporting because it was an area where accurate news was not always easy to obtain and where rumour, fantasy and wishful thinking often obscured the truth. To maintain such a policy was, of course, expensive in so vast a country, but it paid off.

The early success of the *Miner* had been based on the silver of Cobalt; gold came to the rescue when silver failed; then copper; later zinc and lead; and then a host of strange new metals which had to be studied as they grew in value and importance. Nickel, which had been in the *Miner's* field from the start as a war metal, blazed in peacetime as new uses were found for it as a result of one of the most vigorous development campaigns in business history. In gold mining, Porcupine and Kirkland Lake had been discovered shortly before the First World War, but it was not until the 20s when labour became plentiful that they came into their own, together with newer prospects at Red Lake, Ontario; in Manitoba; and throughout British Columbia.

The *Miner* spread its wings—literally, for flying was becoming the way to get around fast; for the newsmen, the airplane increased the jump from mine to office from 100 miles in 1915, to 4,000 miles 10 years later. In a single year the staff logged more than 250,000 air miles. Some of these trips were hazardous—most of them in light planes and, in the early days, in patched up "peach baskets" left over from war service. One old-timer on the staff recalled six forced landings in one season of the Quebec rushes. As mining discoveries spread into the Yukon, Labrador, Newfoundland and into the Arctic barren lands, news-getting became even more adventurous. Today, the airplane, the helicopter, the snowmobile, are as common in Canada as street-cars. Yet there are still times when editors and reporters must revert to canoes, outboards, dog-sleighs and snow-shoes to get around new patches of mining country.

Ever since its disclosure of enemy involvement in the collection of flotation royalties, the *Northern Miner* has continued to assist the industry in its public problems; never has it been a mere mirror of events. Like so many others in its field, it has taken part over the years in scores of battles against discriminating taxation and other legislative proposals. When the paper was started in Cobalt there was no income tax, no corporation tax and no sales tax. As the demands of government treasuries have mounted, the *Miner* has played a leading role in the fight to keep them from damaging the industry. To what extent taxation has killed off ore and employment in Canada can never be realistically assessed, but among the "raids" successfully beaten off was one in the 20s by the Ontario Government to raise mine levies, a threat which led to talks of secession by the North. Another was a gold bullion tax, which the Bennett Government in Ottawa tried to impose when the gold price was raised in 1931. Such a tax would have robbed the mines of the benefit of the increase; the fight was a bitter one, but the mines' position was protected.

In the Second World War, the editors of the *Miner,* privy to the secrets of uranium, knew all about the tremendous power possibilities of the atom bomb. They privately urged the Canadian authorities to take over the Eldorado Mine so that flurries in the stock would not unwittingly tip off progress in bomb development. State ownership was no policy of the paper in normal circumstances, its editors believing that governments can never have the spirit and boldness which animates private enterprise, but this was a special case involving the safety of the Allies in a critical period of the war.

The *Northern Miner's* general policy has always been to encourage mine development, but over the years it has found it necessary to draw attention to undesirable situations and, in doing so, it has, like the *Canadian Mining Journal*, under Bell, and London's *Mining Journal*, under Henry English, been involved in numerous libel actions and countless threats of litigation. It was not always that the public appreciated such disclosures. Investors and speculators, for example, were slow to realize that news which could save them money was as important as news which could make them money. In one of its early years the paper published a story which was far from complimentary about a popular British Columbia gold mine. Its ore position was deteriorating, but directors had failed to tell the public. The *Miner's* reward was that all available copies in Vancouver were gathered together and burned in one of the main streets. Many subscriptions were cancelled. But there was a quiet return of western readers, and many new subscribers, when the truth about the mine's position was eventually admitted. Years later, people were offering to pay $5 a copy for the paper during the Timmins rush of 1964, and almost as much during several earlier rushes in Quebec.

When, in the early 1920s, Kirkland Lake developed into an important mining centre, the Pearces could not resist the temptation to get back into the general newspaper business and they started the *Northern News*. At first it was printed with the *Miner* at Cobalt, but later offices and plant were set up at Kirkland Lake. Running a newspaper in a small town is always an adventure, but an unusually serious problem arose during the bitter Kirkland Lake strike in the winter of 1941–42. The paper supported the mine-owners in a series of vigorous editorials by Norman Pearce in his best style, and the Union retaliated by decreeing that the food vouchers it was handing out to their striking members in lieu of strike pay, were not to be cashed at any store that advertised in the *Northern News*. The paper survived, but it was not in very good shape by the time the strike ended three months later. When an offer to purchase came along in 1946 from the publisher of the Timmins Press, the owners jumped at the chance. This was the first acquisition in the huge international newspaper chain subsequently assembled by Lord Thomson of Fleet: the Pearces were later to reflect on how proud they were of the opportunity of giving him a hand up the ladder!

Since then the *Northern Miner* company has concentrated on developing a high-quality printing house in Toronto. In 1970 it acquired a magazine concerned with broadcasting, which it had previously printed for nearly 20 years, and in 1974 it took over the Vancouver company which had published the monthly *Western Miner* for nearly half a century.

The *Western Miner*, produced in the more traditional magazine format, began life as the *British Columbia Miner* in the late 1920s, at a time when the Sullivan mine had pushed the province into an important position in the Canadian industry. The publisher was Gordon Black, who remained president of his company until his death in 1977. His partner from 1946 was C.H. Mitchell, whom he had hired as editor in 1939 and who returned after spending much of the Second World War as a prisoner of the enemy. Norman Benson joined the partnership in 1956 and he and Mitchell edited the magazine until its take-over by *Northern Miner*.

The first issue in 1928 had been a respectable size of 44 pages, with some 26 pages of advertising, but things became difficult in the great depression which hit British Columbia hard a couple of years later and resulted in the closure of most of the base metal mines and half-time working at Sullivan. Somehow the magazine staggered through the bad times, fared well enough during the War, and then surged ahead as the mining industry rose to the demand for raw material for all kinds of consumer and capital goods. The only casualty was the province's gold mining, which remained depressed for many years and has never really recovered, despite the rising price in the 70s. The paper endeavoured to take advantage of the oil boom in the mid-50s and added to its title to indicate its coverage of the new industry, but it could not meet the competition of new papers devoted entirely to the interests of oil and gas and soon reverted to its old role of covering almost exclusively the metal mining scene. More recently, the increasing demand—mainly from Japan—for high-quality metallurgical coal, which Western Canada has in abundance, led to the company producing a quarterly tabloid, *The Coal Miner*, which is distributed with the *Western Miner*.

The *Western Miner* has never claimed any spectacular *extra mural* activities; it is—and always has been—a well-produced, well-written magazine, directed mainly to those whose mining interests lay westwards from Manitoba and, more especially, within the borders of British Columbia, though copies are to be found in most Canadian mining camps wherever they may be. The paper voiced much concern in 1952 when a new government proposed punitive taxation on the export of iron ore—at a time when Japan was showing interest in becoming an important customer. Mitchell took a leading part in crusading against the government's proposals. Victory was won and the taxes were modified—but it was all an anti-climax when the demand for the product disappeared!

More recently, the *Western Miner* has again taken up the cudgels on behalf of the industry against an array of discriminatory taxation and royalties introduced by government to extract the maximum possible revenue from metal producers. This is, of course, a battle which has no frontiers in Canada—or in other parts of the globe—and will doubtless continue until the world's bureaucrats eventually learn the wisdom of the fable of the goose and the golden egg.

CHAPTER THIRTEEN

Rescue Below Zero

When, in 1927, plans were first made for large-scale exploration of the mineral possibilities of the Arctic and sub-Arctic, Norman and Richard Pearce, co-editors of the *Northern Miner*, sensed the possibility of an exciting story—exciting, even for Canada, where most big mining developments had been preceded by romantic overtures.

For years the brothers had been reading of trips of explorers, of mineral discoveries dating back to the 17th century, of great areas of ideal geological conditions. The mining potential of the Far North, coupled with the possibility of the development of a major south-north railway system and the importance it would have to the future of Canada, fascinated them. They took a keen interest in the preparations of the two principal companies which were formed to undertake exploration of the desolate North-West Territories—Jack Hammell's Northern Aerial Mineral Explorers (NAME) and Colonel C.D.H. MacAlpine's Dominion Explorers. The companies chartered schooners in Nova Scotia and engaged captains and crews to spot caches along the shores of Hudson Bay and the Arctic coast. Together, they employed several hundred prospectors who found most of the showings that have since been worked, including nickel at Rankin Inlet, and zinc at Bathurst Norseman Mines. NAME also did work nearer home and its prospecting teams found Pickle Crow, which became a successful undertaking and a good dividend-payer.

In the late summer of 1928, Colonel MacAlpine invited Richard Pearce to accompany him on an inspection by airplane of the sub-

Arctic bases which his company had established. The 4,000-mile trip was probably the most ambitious journey undertaken by air in Canada up to that time. It took 40 flying hours, spread over 12 days, and the little aircraft became the first to cross the Barren Lands that lie between Hudson Bay and the Mackenzie River basin. Richard returned impressed with the vastness of the country, and with a wholesome respect for the early explorers who travelled parts of it on land and water; he also realized how essential it was for modern prospecting parties to be properly equipped and well organized, for the country appeared to live up to its reputation for cruelty, even to the few who lived there and thought they had begun to understand its moods. Finally, he was able to debunk some sensational newspaper stories of a "race for riches" by the various prospecting teams, between whom he found a spirit of co-operation rather than any deep rivalry.

A few months later, Norman Pearce made a similar trip and in the early fall of 1929 Richard was again asked to join Colonel MacAlpine on a much more ambitious journey of some 20,000 miles. This time it was planned to cover Hudson Bay, the western Arctic coast, Mackenzie River, the mountain country to the west, and the Great Bear and Great Slave Lakes. It was an opportunity that no news-paperman could possibly refuse—to be one of a party that would see more of Canada than had ever been exposed in a single expedition, and to obtain first-hand knowledge of an enormous sweep of territory possessed by no other men.

Richard fully appreciated the hazards of such a trip, despite all the measures which were being taken to ensure an expeditious and safe journey. But neither he, nor anyone else, could have imagined that what had been planned to occupy a few weeks would drag into months, and that the adventures of a cold and hungry group, marooned in ice and snow within the Arctic Circle, and the attempts to find and rescue them in the biggest air search yet undertaken, would provide a story that would ring round the world.

The full complement of the expedition was:

Lt-Colonel MacAlpine, *Dominion Explorers*
Major G.A. Thompson, *pilot*

Donald Goodwin, *mechanic*
Stanley MacMillan, *pilot*
Alex Milne, *mechanic*
E.A. Broadway, *pilot and mining engineer*
Major Robert Baker, *Dominion Explorers*
Richard Pearce, *editor, Northern Miner*

The party (with the exception of Major Baker, who joined at Baker Lake) took off in two aircraft, equipped with both floats and skis, from Winnipeg on August 24 and reached Churchill, on the western shore of Hudson Bay, without incident four days later. There, trouble started almost immediately. It had been arranged that the expedition should be met by the company's schooner, *Morso*, but there was no sign of her, and radio signals were unanswered. One of the aircraft was despatched to search, but returned after sighting a big ice-flow, but no ship. An anxious night and day followed, but towards evening two ship's boats were sighted a mile or two off the little harbour: they contained the crew of the *Morso*, which had blown up and sank after a fire in the engine room. Two of the crew had been injured but the rest were in fair shape after their ordeal. After making the shipwrecked as comfortable as possible, it was decided to haul one of the aircraft onto the beach overnight, but leave the other at anchor in the harbour so that the two injured crewmen could be flown out at first light without delay for the railhead at Le Pas, 500 miles away on Cedar Lake, where they could be given proper medical treatment and sent home.

The next morning the party awoke to find that the plane had dragged anchor and drifted out to sea; a passing ship spotted it five miles out, with all its wings submerged, and took it in tow. Later an attempt was made to lift it out of the water, but the rope broke and it sank; however, when the ship pulled anchor, the plane was found hooked to it, and it was hauled on deck and brought into harbour in a miserable state that evening.

Incredibly, the next night the other aircraft (AO) drifted out to sea and was found floating undamaged by a tug, whose skipper returned it to its owners with some uncomplimentary remarks about people who couldn't look after their airplanes. When a relief plane (SK) arrived a day or two later some members of the party sat up all night, letting it ride out with the ebb tide and hauling it back at the flow.

On September 7 the flight continued to Chesterfield, at the north-western end of the Bay, and then on to Baker Lake, not far from the end of the long Chesterfield Inlet. From there it was hoped to make a 400-mile hop to the prospecting base at Bathurst in one day, but frost in the carburettor of aircraft SK delayed the start until late afternoon and only a short flight was possible to Beverley Lake. From then on, the story is best told by extracts from the diary which Richard Pearce kept throughout the trip:

September 9

> There was fair visibility when the planes took off at 10.10 a.m. Because of the unreliability of the magnetic compasses, sun compasses were used. Pelly Lake was passed over at 11.30 and the plan to keep north so that Bathurst wouldn't be missed was followed. A large lake, bigger than Pelly, was followed for nearly an hour and shortly afterwards we got into violent storms. Eventually we were able to get down on a little lake at 2.50 p.m. A big change in temperature was met and a little snow was passed through. Ice, probably from last winter, was seen filling deep valleys along rivers and creeks. Left the unknown lake at 3.20, each plane with over two hours' gas, according to gauges. Visibility soon became spotty again, but eventually a big body of water was noticed on our right with an ice flow on the horizon. The planes swung along the coast line in a north-westerly direction, the first assumption being that a big bay in Bathurst Inlet had been reached. Near the mouth of a river a building was seen, but it appeared to be an abandoned post. Thirty minutes later an Eskimo camp was spotted and the planes landed at 4.50 p.m. A tent was pitched and camps made. AO reported in excess of 16 gallons of gas and SK 40 gallons. Eskimos visited us, an old man, a young man and wife and baby.

September 10

> The weather was poor today and impossible to fly. Following a good confab with the Eskimos the decision was made to put all the gas into one plane and send it to the post for more. We

found a fine lake, about a mile from the camp, good clear water, and our preliminary exploring of our surroundings indicate that we have chosen to settle on an island. Joe, as we have named the young Eskimo, brought us dried fish, which we tried with sad effects on Colonel MacAlpine. Thompson and I are sleeping in SK, MacMillan and Milne in AO, and the others in a tent. The party is starting to wish that cigarettes and tobacco had not been given so freely to the men at Baker Lake. Joe and his wife are steady smokers.

September 11

It snowed during the night and ice is starting to form on the small lakes. Winter cowling was put on SK in preparation for its flight. The motor was tested out and found to be working O.K. Joe agreed to go with the plane and stayed overnight. The weather continued poor, with a north-west wind continuing. That was the direction to the post indicated by Joe. I shot two ducks and Broadway got a couple of ptarmigan. Made a stew, but the ducks are not very edible. Pearce did the cooking, which might partly account for it.

September 12

All preparations were completed for the SK's trip to the post. Sixteen gallons of gas were taken from AO and put in SK. The north-west wind continued, but with Joe's assurance that it was only a short distance to the post, it was decided to wait no longer. SK warmed up and taxied out in the morning, but trouble again developed in the carburretor intake, so SK came in. All but 9½ gallons of gas had been used. This was transferred back into AO, and with good visibility she started off with MacMillan, Milne, Baker and Joe on board. They returned a few minutes later after finding it impossible to make the crossing of the sea. A conference was held and it was decided that it was best to prepare at once for winter, rather than chance a trip before the freeze-up. It was decided that Major Baker, who was on his way to Bathurst to take charge of the Dominion Explorers base there, would be in charge of the party, and in case of differences of opinion, that Col. MacAlpine would be final arbitrator. All of us realized our position, but the morale was good. The chief worry was not personal inconvenience, but home conditions and anxiety that we would cause our families. Joe brought more salmon, and we are becoming used to it.

September 13

As if to mock us, we woke up to a fine day with the wind
blowing from the north-east. Joe told us a storm was coming,
only a small one, but we could soon expect a lot of snow. The
ice is forming rapidly on the lakes now. We started on a house.
It is proposed to build it of stone, mud and moss and install a
small stove made from a plane engine cowling. The roof is to be
the wings of the AO. We made a good start on the house, but
we found it tiring work gathering rocks and sod.

Joe, his wife and baby moved down alongside us. Joe is very
amused with our house, but it would be difficult to carry on
much longer with a tent and some of us sleeping in the planes.
We got 11 ptarmigan and two ground squirrels; the latter, Joe
says, are good eating. I took stock of all food and ammunition
and Major Baker decided on two meals daily. We had one at 8
and the other at 3. Tea was served at 9. It thawed a bit in the
afternoon and most of the snow went. However, when I went
down to the lake for water at 7 p.m. there was an inch of ice
formed since 3. It was proposed tonight that two or three men
go by canoe to Cambridge Bay, but decided not to do so when
Joe refused to go along.

September 14

Progress was fairly good on the house today, the walls being
brought up to about three feet. We were forced to lay off at
times because of snowstorms. Broadway got three ducks, which
have gone into the stew with two ptarmigan. Joe and his family
visited us after we had moved our tent to a new site near theirs.
Our first choice of campsite was wrong, the melting soil starting
to ooze water after the third day. The tent is now on fine rock
of a broken boulder. The Colonel is starting to lose weight fast
and had Alex put an extra hole in his belt. The rest of the party
seem to be doing well. The wind has again switched to the
north-west, which has proved the wisdom of yesterday's decision
not to try to go to the post by canoe. Everyone is very tired
tonight. I gave Joe four birds for a nice salmon. The boys are
starting to save cigarette butts to use in pipes. A new stove was
built in the tent to save oil and gasoline. The Eskimos showed
us what moss to use with little twigs from small bushes, so we
started the stove.

The boys saw two big swans today and when I went for water
at dusk a big white owl flew over me. Another Eskimo family

came down the river today. We decided to name this place
Dishwater Point. The reason follows: The ducks we got were
not regular ducks, but what we used to call hell-divers, and
when I stewed them a thick layer of grease scum formed. We
all got sick and after we had recovered figured it was a good
time to have a rum ration from our small supply. I put on
water to heat and when Bob was mixing the ration he picked up
in the dark the pot that the greasy dishes had been washed in.
The rum tasted queer and the Colonel asked Bob for a drink of
hot water. He got more dishwater and vomited immediately.

September 15

House building is going ahead at a fast rate, which is a good
thing, because it now seems sure that we will have to stay here
until freeze-up. The walls of the structure are up four feet and
we hope to have the roof on to-morrow.

This has been a tiring day for all of us, carrying frozen sod for
the mud house and gathering fuel. Eight or nine hours' hard
going, and only two scanty meals, could hardly be considered a
satisfactory day.

It turns out that the new Eskimo family is in some way related
to Joe. There are two children, 5 and 3. With Joe's father, they
visited us at noon, making 16 people in a seven by nine tent.
The latest Eskimo arrival, whom we have named Jack,
reiterated what Joe has told us about the impossibility of
making the post now. We offered a whaleboat and a big store
of supplies if he would take out a message, an offer which he
refused.

September 16

We hoped to be sleeping in our stone and mud house to-night,
but it will be impossible to get the wings off plane AO until the
ship is turned around at some high tide. Measured the house today for curiosity and it is about five feet high, 12 by 14 inside
and three-foot walls, with rock as well as muck used on the
lower half.

Joe talks of caribou and we are hopeful that we get some. Our
food supply was helped out a bit when Jack brought a bag of
dried whitefish. We plan to stay up late tonight to swing AO
around so we can take off the wings tomorrow. There has been
speculative talk in idle moments as to whether the Bathurst

wireless station is working. If it is, we hope an alarm has been sent out and planes will soon be looking for us.

September 17

This is the brightest day we have had since leaving Winnipeg. It is quite warm and the snow will probably go quickly. The sea is calm and there is very little wind. We figured this was a favourable time to light a smoke fire on one of the grass flats. This was done, in the hope of attracting the attention of anyone who might be looking for us. A wind from the east came up and carried the smoke along the ground rather than straight up, as at first.

September 18

Mac is bushed all right, so Bob (Baker) says, because when he was up the river hunting Jack called him into his tent for some tea and Mac tried the huskie way of eating dried whitefish, which is to alternate bites of raw fish and caribou fat. I suppose we will all be doing that soon. We have tried the raw salmon and it is not so bad, so perhaps raw whitefish will be fairly good eating. Our fishing efforts so far have been a dud, not a single one being caught.

Decided not to wait longer for favourable conditions to swing AO, to take off wings, and got Joe's wife to sew two tarpaulins together for a roof for the mud shack. It worked all right with a ridge pole made from struts taken off AO.

We moved into the mud house and it is quite an improvement over sleeping in the tent and planes. The floor is going to be tough to sleep on, being of rock, but it should at least be dry. Joe just came in, saw an orange and said he would like it for the baby. He got it.

The boys are talking about the probability of a plane coming in here for it is pretty well agreed a search for us is now on. Everyone in the party feels keenly our helplessness and inability to move toward a post at this time. Joe told us again to-night by illustration that a canoe would be swamped, and I guess he is right. At any rate, the Colonel is positive that no one take a chance on a move that the Eskimos will not take part in, though all of us have offered to do so, if he would give his consent.

Jack moved his family down near us; the total: a wife, two kids and seven dogs. The dogs had packs on their backs, but carried

little more than 25 pounds each. The more company we have
the better. It is much colder to-day, and we had the gas torch
going in the shack, trying to dry out the mud walls. It is doing
the trick and we should get more sleep to-night, if we are free
from the clammy dampness we have had so far.

At supper time all the Eskimos visited us. A stew of two
ptarmigan and a pot of tea fed sixteen. Mac, Alex. and I went
out for a few miles in the afternoon and got three ptarmigan;
scared up an Arctic hare, but didn't get him. The hare is much
larger than any rabbit I have seen before and looks to be nearly
the size of an Arctic fox.

We showed Joe and Jack a flare to-night and tried to persuade
them to go to the point near Kent Peninsula, opposite
Cambridge Bay, and light off flares there. They refused point
blank and pictured a canoe swamping, and said it was
impossible to go overland.

September 19

Our whole party went out on a hunting trip at 9.30 and got
back at 3. We spread out so as to make sure of scaring up
game if it was to be found. It was tiring work but Joe came to
the rescue with a pot of tea on our return. Our stew came along
a little later and, in time for it, the two Eskimo families.

Some of our boys cannot forget that they are hungry, so the
favourite topic of conversation is how much they are going to
eat when the opportunity presents itself, or the goodness of
steak and onions or some other fine dish now out of reach.

Jack brought in slices of caribou fat for us all and it tasted
pretty good. The Colonel downed his all right but said part of it
was awful. The part he didn't like was the outside edge that
had turned green. It was probably a year old.

The inland lakes and rivers are pretty well frozen. Our trip on
the island to-day showed it to be much different from the
Hudson Bay barren lands. There is an abundance of tiny
willows. Grass-covered flats that feel underfoot like the muskegs
of Northern Ontario are found between the rock ridges. The
flats are frozen and make good walking, but the rock ridges are
hard going.

September 20

Two young Eskimo men and a young woman, presumably
brothers and sister, came in to-day. The newcomers paid us a

visit along with Joe, Jack and their families. There are now 12
Eskimos with us and we figure at least one more family remains
up the river, as there are five dog sleds in the caches near our
camp. AO is frozen in and if the wind drops there should be
quite a lot of ice formed on the sea by morning. The ice on our
fresh water lake is now all of four inches thick.

And so life continued at "Dishwater Point", or Dease Point as it is
named on the map, near the open shore of Queen Maud Gulf, opposite
Victoria Island, and the post at Cambridge Bay, so near and yet so
far, where the marooned men knew a Hudson's Bay Company ship
was wintering, where there was warmth and shelter—and radio to
enable rescue plans to be put in operation and to inform their families
of their safety.

For Major Baker, who by unanimous consent had assumed command
of the party because of his knowledge of the terrain and its changeable
moods, there appeared to be three alternative courses of action: first,
to remain where they were until the waters in the Gulf froze and they
could travel over the ice by dog-sled and on foot to Cambridge Bay;
second, to try to get to the post by boat; third, to move to a point
nearer the shore and try to signal the post by flares.

The Eskimos ridiculed the idea of trying to go by boat; in any case
there was only one canoe and that was quite unseaworthy; they made
it clear that they were not prepared to take the risk themselves,
declaring that the boat would soon sink—as it undoubtedly would
have done. Earlier in the season it would have been feasible to move
closer to the Gulf shore opposite Cambridge Bay and try to attract
attention by flares, but it was now much too dangerous: the small
lakes had started to freeze the day after the planes' arrival and every
few days there were blizzards and cold rains. Even for this trip, a
good boat was essential as there were rivers to cross; furthermore, as
the Eskimos pointed out, the operation would have been pretty useless
as Cambridge Bay was surrounded by high hills, and signals from
across the sea would probably never have been seen. The only way,
therefore, was to wait until the sea had frozen hard enough to travel
across the Gulf on the ice. It was obvious that this was the way the
Eskimos saw it—and the way in which they were prepared to help.

On September 21, one of the Eskimos raised the party's hopes by drawing a map showing how it was proposed to get to Cambridge Bay once the ice was hard, and where it was planned to stop overnight. This was the first definite indication that the Eskimos had taken matters into their own hands and decided to lead the party to safety. Sensibly, the explorers accepted this. But it was to be a long wait. The days came and went, and nothing seemed to happen. There was great consternation when, after supper one evening, it was found that the Eskimos had broken camp; hopes of rescue plunged, but it was soon discovered that they had only left for a hunting and fishing trip to re-stock the larder. September faded into October; the winds got stronger and colder and, with the approach of winter, the ice began to thicken:

October 15

> We took a trip among the islands and saw such a maze of them that it would be impossible for us to find our way through them without a guide. Last night was real cold and no less than four inches of new ice formed on the water hole. We are hopeful this means the closing of the straits.

October 17

> A gale started up last night, blowing from the south-east, and gloom settled on the party, for fear the wind would break up the ice in the strait. The weather being bad, it was no fun out, and we decided to stay in bed. I got tired of bed and got up and had no sooner done so than I heard dogs. I kept quiet until I was sure I was not dreaming and then gave the alarm. A few seconds later our shack was filled with Eskimos. We gave a joyous welcome to our old friends. They kept straggling in until 4 o'clock and eventually 16 arrived and were fed. With 24 in our little shack it was no easy task, particularly as some of the youngsters kept wandering around.
>
> The gale increased and the roof of our shack gave signs of leaving us. The blizzard is now beyond anything we have seen before and it is impossible to see five yards ahead. How the Eskimos ever made the trip today is beyond us. They were snow-covered from head to feet, but even the little tots took it as a matter of course. They told us the straits were frozen and that they would take us to the post as soon as the storm died

down. This news was as welcome as if someone had told us that we had been left fortunes unexpectedly.

Four days later, the Eskimos announced that it was time to leave. The party had been at this desolate place for 41 days. They had 80 miles to go, mostly along the frozen shoreline, with the last 25 miles a dash across the ice to Cambridge Bay. In the first day they covered more than 20 miles from the mud shack that had been their home, quickly discovering that dog-team travelling was not like it was on the movies, with sleds speeding effortlessly across the snow: the reality was slogging footwork, especially when the snow was soft. They slept in igloos which the Eskimos soon had erected with slabs of ice—and which caved in with the warmer weather at daybreak, when they were off again, but at a slower pace, covering only about 10 miles before again making camp. The next day, warmer conditions and resultant slush prevented any movement, but the following morning they got going again and made good time until they struck ice hummocks which caused more delays—and two more uncomfortable nights with little sleep.

At last, they reached the point at which they were to cross the frozen strait—only to find, when they were a quarter of the way across, that tide and wind had combined to produce a break, 100 yards or so wide, between them and the distant shore. There was nothing to do but turn back. For six more days they waited for better conditions; patience was near breaking point, and the first rumblings of mutiny occurred when some members of the party announced their intention of pushing on regardless of the warnings of the Eskimos. But reason eventually prevailed and November 2 proved to be D-Day—the dash across the ice of the Dease Strait.

Pearce's diary tells the story of their crossing:

November 2

    It is early in the morning, and there is a conference of Eskimos in our igloo. They appear to be talking over the trip today. We had a wonderful breakfast of three cups of tea, a half bannock and a big slice of boiled Barren Land trout. Zero hour approaches, and I hope we spend the night at Cambridge Bay.

(Later) We spent hour after hour trying to pick a course through a large ice floe, wandering this way and that to find the best going. Sleds were tugged and pushed over ice hummocks that we uninitiated thought were impossible to cross. More than once we tumbled into holes between ice cakes. It was all nerve-wracking and tiring.

Narrow leads of unsafe ice had to be crossed, but before we ventured on them the Eskimos tested them with spears. Alice, the wife of the Eskimo with whom the Colonel, Baker and I were travelling, took her boy of three after he had been tossed from the sled several times, stripped him naked and tied him on her back under her clothing. This did not stop her from continuing a man's share of the work of urging the dogs and helping tug the sleds over the most difficult spots.

Once, while crossing weak ice, apparently frozen only the night before, she broke through to her knees. Instead of trying to scramble out—which would probably have meant that she would sink deeper—she lay flat until one of the Eskimos rescued her. On we went for a quarter of a mile, when Alice changed into dry clothing out in the open without any apparent great discomfort, although the temperature was well below zero.

At 4 p.m. we struck thin ice. The Eskimos could find no way to cross it. It was then a matter of parking for the night on an ice floe, hoping that in the morning the going would be safe. Some of the party feared that the strong wind then blowing would carry the floe down the straits and perhaps break it up. The Eskimos assured us, however, that the wind would die down during the night.

The Colonel, Bob Baker and I crawled into our sleeping bags. I don't think I was ever so tired before in my life. I had to be wakened between cups of tea. The Colonel was in agony from chapping and from the seemingly eternal lifting of feet, but there was never a complaint from him.

Throbbing pains in my shoulders kept me from sleeping long. No doubt they were caused by the wrenching from the sleds. As the Eskimos are still up I have an opportunity to bring the diary up to date . . .

November 3

This is the night of nights. We are at Cambridge Bay.
Messages have gone out to our families, and we can go to bed
in peace. But what a day! It started at 5 a.m. with a very scanty
breakfast and then a race across thin ice. The Colonel, who was
in great discomfort, got a ride part way over the first five-mile
stretch, but the rest of us had to run, dodging the dark green
spots in the ice that we could see wavering under the men
ahead.

The dogs were urged by word and lash to keep moving. My
second wind passed by without recognition and I was running
on my fifth, about the second mile or so. How we kept going
I don't know. We simply had to. We knew that if anything
happened to one of us, there would be real danger in others
stopping to try to help.

By the time we got on safe ice it was a weary lot of travellers
who followed the sleds. Our legs were tired, sore and chapped,
and the bitter wind with a temperature of 27 below zero bit
into our lungs. Toes, faces and fingers were frozen. The thin ice
had a layer of salty slush on it which worked into our footwear
and in no time moccasins, duffles and socks were frozen
together.

Poor Don Goodwin! I can see his face now as he trudged along
with feet frost-bitten from the day before. He apparently didn't
think it was a serious matter, because he didn't tell any of us
last night. As we travelled, and one member of the party came
close to another, a hurried search was made for frozen features.
Four times my nose was thawed out during the day. My
personal casualty list for the day was six finger tips, two toe
tips, both cheeks and nose frost-bitten. This was probably an
average for the rest of the party except Don, who was the worst
hit.

We had figured we had 12 miles to go, but we must have
travelled at least 30, dodging around the floe ice to get the best
going. Just when we needed a rest most, the dogs played out.
After that it was a continual walk or trot as the going
permitted.

None of us will ever forget November 3, 1929. Miles away
from Cambridge Bay—and the miles seemed extra long—we
could see what looked like a beacon post for ships. It turned out
to be a mast of the *Bay Maud*, wintering in Cambridge
Harbour.

No more wearied travellers than we were ever got into
Cambridge Bay, and what a splendid welcome we got! Hudson's
Bay men, headed by the manager, police and Canalaska trading
staff could not do too much for us. Since morning, all we had
had was a half-cup of Oxo, and when someone produced a
bottle of whiskey my share was passed. I started to faint and
asked for water. Someone passed more whiskey, and I passed
out for five minutes.

Meals were soon produced. All of us have eaten too much, but
we had potatoes, jam, marmalade, bread, cake and cheese, and,
as someone said afterwards, it was worth getting sick. Don was
hastily fixed up, his footwear being cut from his feet.

The *Bay Maud*, with its wireless masts, was the first place the
Colonel struck for. What a joy it was to pound out words on a
typewriter, words that would be wafted south to our families,
even though every tap of the keys sent a tingle through frost-
bitten fingers. I could easily have cried as I typed a short
message to my wife.

After midnight came the first real bed since August 25, with
something soft underneath the body and a covering of dry,
clean blankets. Just before going to bed the Colonel, Bob and
I cleaned up three tins of tomatoes. Mat Shand, in charge of
the *Bay Maud*, told us that all the Colonel had eaten for supper
was six eggs, a liberal helping of bacon, bread and other things,
and three-quarters of a raisin pie. His second meal at the
Hudson's Bay house later in the evening was equally "modest".

Half-way through the night, Baker and I awoke, had another
meal, and to help pass the time, this diary was resumed. Then
I climbed into Bob's bunk and we ate a pound of chocolates
between us.

This should have been the end of the story, but persistent storm and
blizzard, and misadventure with airplanes kept the survivors marooned
in the Arctic for another month before the first of them could be flown
out to Le Pas to board a train for home. The search for the missing
had been a huge operation, involving scores of outposts, teams of
Eskimos and endless patrolling by every aircraft which could be
pressed into service. It cost many thousands of dollars and as a result
more than half a dozen planes were stranded in the Barren Lands that

winter. The rescue pilots had failed only by the narrowest of margins to solve the mystery of the party's disappearance: search planes skirted the shores of Queen Maud Gulf and one flight actually crossed the route of the lost aircraft. As it was, the missing men had to battle their own way out, guided and helped by their Eskimo friends, famished and freezing in the bitter gales of the Arctic coast.

For Richard Pearce, there was plenty of work during the enforced stay at Cambridge Bay, for the world's Press was clamouring for the story of the rescue. It was decided to write one full account and sell it to a syndicate for $10,000 and the money was divided among those who had suffered the ordeal and the rescuers.

Whether the experience of the MacAlpine party in 1929 had any direct effect on the pace of mineral development is, of course, questionable. However, the efforts of its members and their observations, and the work of the rescue planes have since been considered to have advanced Canada's front line in the north by at least 25 years. If nothing else, they showed that the Arctic, through planes and wireless, and the spirit of determined men, could be conquered. And Richard Pearce, of the *Northern Miner*, was the man who told that story to the world.

The Challenge of Change

Excluding those papers whose interests are largely confined to metal-marketing—and which are reviewed in Chapters 16 and 17—the most influential periodicals in the mining world in the first half of the present century were *Engineering and Mining Journal*, of New York, and *Mining Journal*, of London.

E/MJ's "years of chaos" in the early 1900s, when the paper changed hands four times in as many years, gradually became stabilized under the ownership of John Hill and the editorship of Walter Ingalls, who accepted the post after T.A. Rickard's brief, but traumatic, experience, on condition that he had complete editorial control—and was permitted also to serve as a consultant for the United States zinc industry. Ingalls raised the standard of content by including contributions by some of the most eminent mining men of the day, including Herbert Hoover, who wrote many articles of international interest, including an outstanding review of gold mining in Western Australia.

In its early days, the magazine had paid considerable attention to coal, largely because of Rothwell's personal association with the coal mining industry and his enthusiasm for collecting and publishing statistics of its production in the United States. After Rothwell's death there was not much editorial interest in the subject, and coal mining men began to complain. As a result, Ingalls added a coal expert, Floyd W. Parsons, to his staff and, as well as including news and features in each issue, he initiated a special number entirely devoted to coal every July. Ingalls was always careful to keep a good balance in his paper, allocating editorial space strictly in proportion to the number of subscribers in each field. This gave those with coal interests about one-eighth of the total space. But still they were not satisfied, so Ingalls persuaded Hill to start *Coal Age* as a separate publication: it was an immediate success from its inauguration in 1911 and soon had 6,000 names on its subscription list.

James McGraw, who had bought *E/MJ* in September 1901 (for a reported $183,000) and sold it to William Johnston 15 months later (for a reported $285,000), merged his publishing interests with Hill in 1917 to establish the powerful McGraw-Hill Publishing Company, and *E/MJ* became one of a large family, but fortunately with little loss to its character and independent editorial identity.

Arthur W. Allen, who was editor from 1927 to 1933, was chiefly responsible for two new publications within the group: *E/MJ Metal and Mineral Markets*, a weekly containing price quotations and other market news (see Chapter 17) and *Engineering and Mining World*, which represented the first major attempt by the company to move into the international field with its mining publications. The magazine was a monthly aimed specifically at readers outside the United States, and concentrated editorially on the operating, technical and business problems of producing and processing metal and non-metallic minerals. Despite a circulation of more than 6,000, it lasted only a couple of years as a separate entity before being merged with the parent publication.

E/MJ owes much to the lessons which it learned from James McGraw, one of the most successful industrial publishers of all time. McGraw taught that foremost among the ideals of industrial publishing was that a journal must be its own master, having no other guide for its opinion or policies than truth and a sincere interest in the field that it serves. Many men, warned McGraw, will try to bend the policy of a paper to serve their own selfish interests, but the publisher who stands fast, who protects the integrity of his editorial pages, will have the satisfaction of knowing that he has kept faith with his conscience and his readers, and that those whom he has served will, by their support and confidence, crown his efforts with success. McGraw insisted that the interest of readers must always be paramount and, in the *E/MJ* context, the prime objective should be to give them the kind of information that they needed in order to do a better job producing metals and minerals for society. "An editor", he declared, "must be a great teacher."

Among those who accepted the McGraw philosophy without question was Alvin W. Knoerr, editor of *E/MJ* from 1955 to 1969, since when he has continued to be engaged in editorial work at the U.S. Bureau of Mines in Washington. Knoerr did not need to be told about the value of teaching as he had worked as an instructor at the Wisconsin School of Technology, formerly the Wisconsin Mining School, at which he himself had graduated. Moreover, before he started with McGraw-Hill as an assistant editor in 1944, he had had a varied career in practical mining and his appointment as chief editor of *E/MJ* heralded a new era in the magazine's history.

Knoerr immediately demonstrated a natural flair for the job and a sense of immediacy. His own philosophy embraced everything that McGraw had taught, together with many ideas of his own. His policy was based more on examining the problems of the future than on reviewing the past, or even on assessing matters of current concern. When the paper reached its centenary in 1966 he devoted the whole of the celebratory issue to exploring the challenges of the future rather than following the traditional practice of indulging in the pleasures of history.

"The need for such appraisal", he wrote, "was suggested by the accelerating technical revolution under way today, and the intensified evolution of a political world in which mining will have to orient itself. Of all the challenges reviewed in this issue, the challenge of change appears to be the most demanding. To ignore it with complacency will be equivalent to obsolescence and decay; to meet it will require vitality and ingenuity ... the challenge of change will be as important to mining communications as it will be to the industry."

Those who have had personal association with the industry, particularly in Third World countries, over the past two decades, will appreciate just how prophetic these words were. Failure to recognize the pressures of change has led not only to fundamental reconstruction of many great mining enterprises, with corresponding lowering of efficiencies and profits, but also to political chaos, civil war and economic disaster in some areas of the world.

Knoerr's forward-looking approach was reflected in the Journal in many ways. One was its treatment of mining's manpower problem. A series of articles, spread over several issues, set out to shake the industry out of its complacency over a shortage of engineers and skilled technicians which many regarded as temporary and really not worth worrying about. *E/MJ* insisted that it was a long-term problem and something had to be done about it—and the paper put forward constructive ideas as to how it should be tackled. The magazine also examined the potential of a host of the world's lesser-known metals and pointed ways towards their development and marketing. It ran some of the most comprehensive surveys ever written of the world's most important mining countries and, in the process, won the highest awards offered for competition in American journalism.

E/MJ's most significant contribution to mining since the Second World War was probably the part which it played in prodding the United States industry to participate in what was to become one of the biggest booms in its history—the exploitation of U_3O_8, better known as uranium.

The story goes back to August, 1945 when the first atomic bomb was dropped on Hiroshima, causing indescribable destruction, loss of life and terrible mutilation—but heralding the end of the war with Japan. The event was page-one news in the world's newspapers and, in the Allied countries, the announcement broke a bond of silence which had been imposed on all publications. In the United States, even those who had been directly concerned in mining uranium did not know the real purpose of the operation: it was the best-kept secret of the war. Since the early 40s the United States government had been feverishly trying to accumulate every possible ounce of uranium, under conditions of great secrecy. The biggest area of production was the Colorado Plateau, composed of the states of Colorado, Utah, New Mexico and Arizona. Here, an industry for the mining of vanadium was built up—to provide a cloak for the extraction of uranium.

In *E/MJ* the staff had done their homework and soon after the news of Hiroshima an eight-page feature explained how atom-splitting releases energy; how high-power atoms are created and isolated; and what to expect from atomic energy. This article, which was published simultaneously in other McGraw-Hill publications, was the first complete story on atomic energy from uranium to appear in any mining periodical. Later articles told where uranium was being mined and processed throughout the world, pointing out that its production was widespread and abundant—available to many foreign nations.

At this time the Colorado Plateau, over 100,000 square miles in area, contained the second biggest known uranium ore resources in the world. *E/MJ* staff made frequent trips to the area to obtain progress reports and help wherever possible. It was a chaotic situation as the companies that had been producing uranium for all their worth during the war now found that no one appeared to want it. Alvin Knoerr recalls that when he visited the Plateau in 1947 he saw a whole warehouse stacked from floor to ceiling with uranium. An exasperated manager said, "What do I do with it? There's no price set on it . . . no interest in it. We don't even know if we can sell the stuff."

Then, spurred by the Communist threat and the realization that the A-Bomb was something very, very powerful, the government set up its Atomic Energy Commission, formulated a policy and decided to ask private enterprise to get on with the job of producing as much uranium as possible. The response was lukewarm. *E/MJ* sought to help solve the pressing problem of how to make U.S. mining companies aware of the urgent need for them to get into uranium production by publicising the Commission's appeal. Still there was little response. Alvin Knoerr takes up the story:

"By now, industry had a guaranteed market, a guaranteed price, and yet it still did not show any interest. I went out to the Plateau and visited the operations with some of the Atomic Energy Commission men at Grand Junction. We drove into the rugged interior areas and I saw that a lot of the uranium mining was being done by bean farmers and ranchers who were quick to recognize the fact that there was a dollar to be made. They were using some primitive and unsafe methods, but even so they were making a profit. Our first article was *Can Uranium Mining Pay?* and I remember that the concluding statement ran something like this, 'If the bean farmer and rancher in Colorado can go there and make money, it is high time that the mining industry, which should know how to do this work, did it as well, or better'."

The article had quite an impact—and so did the picture on the cover of the issue in which it appeared. It showed a notice-board, in the middle of the desert, hundreds of miles from the nearest town, reading "Wanted—Uranium Leases". Alvin Knoerr had taken the picture as he drove by in his jeep. "It is not often that an editor can say that he helped make a million dollars for a man, but that's exactly what happened here", he recalls. "Charles Steen was living in Texas at this time and was thinking of going to Mexico to hunt for gold. Then he got a copy of the Journal with this picture on the cover; he read the article and promptly told his mother that he wasn't going to Mexico anymore—he was going to look for uranium. And he did. After two or three years of very painful experiences, borrowing and begging every nickel that he could and practically starving the family, he struck it rich. He found a deposit measured in tens of millions of dollars. He came to New York and told us his story. By that time he was a multi-millionaire. He came in and the first thing he said was 'I owe *E/MJ* a personal debt of gratitude for having discovered this deposit, and I just wanted to come and say thank you'."

Charlie Steen's discovery led to a prospecting boom. Again, the Atomic Energy Commission asked *E/MJ* for help. A great many people were producing uranium ore but their methods were neither skilful nor safe, and in some cases they were wasteful because of "dirty mining." *E/MJ* went back to Colorado and researched the facts for another feature, which pointed up the advantages of mechanical mining and described in detail how the operations could be carried out more economically and efficiently. Later, because of mounting requests for information, the company produced its U_3O_8 *Formula for Profits*, which sold more than 16,000 reprints in record time and remains a classic of its kind. Charlie Steen was only one of many who were inspired by the profit motive which *E/MJ* underlined in its articles, and soon there were more than 550 uranium producers on the Colorado Plateau. Today this is a strong and vital industry, based on private operations; in 1977, more than 10 million tons of ore were produced, containing 16,000 tons of U_3O_8.

For its treatment of the uranium story, *E/MJ* added two more awards to its professional trophies, but its principal reward was the satisfaction of knowing that it had helped its government and its industry in a critical problem and demonstrated the positive part which a paper can play in both industrial and national development.

*

Many successful men owe much to the help and encouragement given them in their careers by their wives. Few, however, could ever have received so much practical assistance from their life-partner over so long a period as Edward Baliol Scott, editor of *Mining Journal* from 1901 until 1954. For 40 of those years, his wife Gladys was closely identified with the Journal's business direction and her work played no small part in the company's successful emergence from the difficulties of two world wars and periods of economic depression, particularly in the late 20s and early 30s.

Gladys Agnes Grant was born at Dovercourt, Essex, in 1879, the only daughter of William Charles Perry Grant, Paymaster-in-Chief of the Royal Navy, and his wife Helen, daughter of Captain James Napier, RN. Educated privately in an era when this was still the conventional preparation of a young lady of gentle birth for the responsibilities of adult life, Gladys Grant took a grip on these responsibilities at an early age. Following her mother's death, she was keeping house for her father at the age of 14 and travelled extensively with him throughout Europe. During much of this time they were in Florence, where she developed a love of painting which was to be her main relaxation throughout her life. Gladys was a woman of tremendous energy and determination, with a flair for organization, both in her business life and in her varied social and civic activities. An outstanding rifle shot, she was a familiar figure at Bisley in the decade before the First World War. She founded the Surrey Ladies' Rifle Association and did much to promote the formation of women's rifle clubs throughout the country.

When war broke out in 1914 and the call to the Forces had depleted the staff of the Journal, she moved into the office to help carry on her husband's business. She expected to be there for six months (for the war was to be over by Christmas, as everyone knew), but she remained for 40 years. With no previous experience of publishing, she learned the hard way, rapidly establishing herself as the company's business manager at a time when a woman in an executive position in publishing, or indeed, anywhere in the City, was still a rarity.

During the '30s she was a prominent member of the Ladies' Forum Club where, on behalf of the Writers' Section, she exercised her particular blend of determination and charm with which, throughout her life, she successfully persuaded people to help her many good causes, and to find, often to their surprise, that they enjoyed doing so. There are probably still a few elderly distinguished journalists and authors in London who will remember with mixed feelings finding themselves a lone male, and committed to speak, at one of the Club's formidable literary dinners.

Having seen most of the Journal's staff vanish in 1914, Gladys was again left to carry on the business side of the publication on the outbreak of the Second World War in 1939. Editorially, the paper owed much during, and in the years immediately after, the war to "Willie" White, who had been the city editor, based in London, of a provincial daily paper.

Although working every day at the Journal, Gladys, now over 60, managed somehow to give part-time service to the Women's Legion through the early years of the war, manning a mobile canteen in London's dockland at a time when it was coming in for some heavy attention from Hitler's bombers. After the war, her energies found fresh outlets in the affairs of the Individual Members' Group of the British Federation of Business and Professional Women. She also entered whole-heartedly into every aspect of village life at Ide Hill, in Kent, which remained her home until a few years before her death in 1965.

Throughout these difficult years, Edward Baliol Scott continued to direct the Journal's editorial affairs, taking every opportunity to widen the scope of the magazine, probing behind the political and economic changes in the mining industry. In the process he became an expert, in his own right, on mining economics, especially in copper, tin and gold. In this, he paralleled the accomplishments of Dr. Joseph Zimmerman, editor of the New York-based *Daily Metal Reporter* for 40 years or more (Chapters 17 and 18), though, unlike Zimmerman, he spoke and lectured publicly only rarely, his philosophy being almost entirely expressed in the columns of his paper.

On occasions he gave close editorial attention to matters that appeared to have no direct relationship with the purpose of the Journal. Thus, throughout 1914–18 he wrote a weekly diary of the war in the conviction that nothing could have a greater interest for the mining fraternity, mostly working in remote areas, than news of the progress of the struggle. As a life-long student of British naval history (and obviously encouraged by his wife, who came from such a distinguished naval background), he wrote with exceptional insight of the war at sea, a service which earned him the personal thanks of Admiral Lord Fisher. He was over 80 when he retired in 1954, the longest serving of the seven editors who had occupied the chair in the Journal's 120 years' history. Even then he did not give up as chairman of the company, and his long experience and advice continued to be available to the editorial staff until his death 10 years later.

When Edward Baliol Scott took over the running of *Mining Journal* in 1901 he had no mining background; he was a classical scholar and barrister, and he assumed control primarily to deal with a management problem and get the business on its feet. Fifty years later, history repeated itself when the company, weakened by the war, again faced management problems and when, at 75, its editor-publisher was beginning to run out of steam. Initially, Ursel, Edward's younger son, had no intention of going into the business. Like his father, he had no mining experience; he joined the company soon after the war from a career in scientific management and wartime experience in aircraft production and stock control.

Despite Edward's success in revitalizing the Journal at the beginning of the century, Ursel held the view that his father had got his priorities wrong in that he tended to concentrate on the editorial side at the expense of the actual business operation—a tendency which is not surprising in view of the young lawyer's background, for the business side of publishing can be deadly dull to someone who has a literary bent. Ursel was determined not to make the same mistake—if, indeed, it had been a mistake—and, come what may, he would redress the imbalance. Twenty years later he was able to look back from a position of solid re-establishment with the success of the company's publications assured. But it was not easy to stick purely to business priorities and he found, like his father, that the pressures and challenge of producing a weekly issue meant that he never had as much time to give to the business side as he would have liked.

However, Ursel enjoyed what he describes as his "editorial period". He occupied the chair for 10 years after his father retired and there is no doubt that in that time he made a considerable contribution to mining thought. He strove to maintain the editorial independence of the Journal, working hard to preserve the reputation his father had developed for editorial discretion without in any sense being syco-phantic. "We write", he once said, "about the mining industry from the standpoint of the people in the industry, but we have never been inhibited from saying what we think is right." He feels that at times he must have annoyed many people, among them—and perhaps more especially—those on whose goodwill the paper depended. "But", as he says, "if you are serving a sectional interest, as we are, you can do it in two ways. You can be, editorially, something of an Uriah Heep, or, as we have sought to be over the years, the industry's 'hair shirt'."

Under the father, in the '20s and '30s, *Mining Journal* often lived up to its reputation as being the industry's "Yellow Peril"; under the son, it was the "hair shirt". It would be difficult for the paper to lay claim to either of those titles today—but that is not to suggest that it is any better or any worse for a change which, over the past 10 years or so, has tended to be towards a higher professional standard—in mining lore, though not necessarily in journalistic skill.

Perhaps the most significant development during Ursel Baliol Scott's management of *Mining Journal* was its acquisition in 1963 of *Mining Magazine* which, after T.A. Rickard's return to the United States in the middle of the First World War had never managed to recapture the spirit of enterprise or the high degree of expertise that he had infused into the project. Perhaps more than most of its contemporaries, it suffered acutely from wartime restrictions and loss of circulation, though it picked up somewhat in the 1930s and regained some of its previous reputation for the technical excellence of its articles. For this it had almost entirely to thank Frank Higham, one of the most kindly personalities in the profession. Higham became assistant editor in 1929 and editor in 1938; he continued to edit the paper after its take-over by *Mining Journal* until his death in 1966.

Like most editors of mining papers, Frank Higham had no early ambitions towards journalism; rather was it thrust upon him by the vicissitudes of a life in which he had more than his share of misfortunes, but for which he could well have become an outstanding figure in the world of economic geology. He began as an articled pupil to an analytical chemist but, as a result of a laboratory accident which cost him the loss of an eye, he turned to geology and graduated from the Royal School of Mines. His first professional appointment—which was also to be his last—was with Nile-Congo Divide Syndicate in the Sudan. There, he was gored by a buffalo and invalided home. During his convalescence he joined the staff of the RSM as a demonstrator in the Department of Geology. He left to join *Mining Magazine* and from 1943 he also edited the *Transactions of the Institution of Mining and Metallurgy* and various other of the Institution's publications.

Mining Publications, which was the company that Herbert Hoover and some of his London friends had set up in 1909 to publish the *Magazine* in competition with the *Journal*, became quite diversified in its activities. It had achieved early recognition in the mining world by its publication of the Hoovers' translation of *De Re Metallica* and, on the strength of this, it started its Technical Bookshop, which was to become known among mining men all over the world. Initially, the Bookshop did just what its name implies—it published and sold technical books for mining engineers and metallurgists, but it soon branched out, offering other services on a "deposit account" basis. These other services were very elastic and the Bookshop specialized in obtaining the "unobtainable" for its clients in distant lands, dealing in everything from sophisticated geological equipment to "naughty nighties" for its younger subscribers to give as presents to their wives or girl-friends. One client even wrote from the Gold Coast to order a costume for a fancy-dress ball. Alas, the Bookshop is no more and its manager and director, E.W. Julian, who was associated with the company for nearly 50 years, is now in retirement in Somerset.

By the early 1960s it had become apparent that *Mining Magazine* could no longer be sustained by itself as a viable publication and, in the words of a joint announcement by Mining Journal Limited and Mining Publications Limited, "it seemed to both our companies that the mining industry would best be served by transferring ownership to Mining Journal Limited, which already publishes the weekly *Mining Journal* and the *Mining Journal Annual Review*."

Mining Journal was quick to disclose details of its plans for the two papers. In the first place, the ailing *Magazine* was to be resuscitated and developed on modern lines, its new policy being to assist technical management from the exploration stage, through mining and mineral processing to marketing. It was to be a well-illustrated, high-quality production, despatched to subscribers each month by surface mail.

For its part, the *Journal* would undergo a change of format, printed on airmail paper and despatched by air to reach readers at the most distant destinations within seven days, and those nearer home much earlier. Its contents would be devoted to up-to-date news about the industry, and to economic and political events and trends affecting its welfare; it would be liberally laced with comment and "interpretative" material. To keep airmail postage weights down to manageable proportions, a major part of its display advertisements would be transferred to *Mining Magazine*.

This package deal constituted a new concept in technical publishing in that for a single subscription the reader received his editorial service in three parts. The rationale was that as the service was to circulate to nearly 140 countries, those elements concerned with perishable news needed airmail distribution. On the other hand, it was clearly uneconomic to airmail the whole service. Consequently, *Mining Magazine* would carry the longer articles as well as most of the technical notes and other features which could stand the delay inherent in distribution by surface mail. *Mining Annual Review* was to continue as a separate publication within the three-part service. This was necessitated by its purpose of providing a continuing year-by-year historical record of events in the industry—country-by-country, metal-by-metal and process-by-process.

There is no question that the venture has been a success. The circulation of the *Journal* before it was "airborne" was around 2,500 and, with the *Magazine*, after weeding out the duplications, the combined total was not more than about 4,000. Within a short time this figure has doubled.

A few years after this reorganization, Ursel Baliol Scott remarked in an interview: "Our treatment of the technical news of the industry and our editorial thinking about technical problems have been very greatly developed. It's very much more professional and expert than it was in my father's day, for there are now five mining engineers on our staff. On the economic and political side, we have tried pretty much to continue the policy developed by my father. I will not say that the nature of our relationship with the industry is quite the same as his. He, I think, was considerably more detached from the industry and the people in it."

Sometimes a degree of detachment is a very good thing when you are sitting in an editorial chair. It must be remembered, too, that any organization that seeks to establish a rapid news service with distant parts of the world must recognize that the success of such a venture depends not only on having something new and constructive to say, but upon the material being written and presented in a manner that demands instant attention and can be easily read. Ideally, mining journalism should be a combination of mining professionalism and journalistic expertise—and it takes as much training and experience to achieve the one as the other. It is a pity that in their preoccupation with striving to secure a high level in the first of these requirements, many publishers of technical magazines often fail to meet the basic needs of the second.

CHAPTER FIFTEEN

Sweet Genevieve

The history of the American West is a history of the endeavours of the small miner, beginning with James Marshall's discovery of gold at Sutter's Mill in California in 1848. The names of the great mines in the United States that have been found by small miners are familiar and impressive: the Comstock Lode in Nevada; the Mesabi Range of Minnesota; the iron country of north Michigan; Arizona's copper; Florida's phosphate; Utah's uranium; and Colorado's molybdenum, silver and gold.

Today, prospecting teams are still looking for minerals in the mountains, in the deserts and in the forests of North America. Many of these expeditions are equipped with the most sophisticated tools and instruments that modern technology can supply. The days of the old prospector with his shovel and pan, his mule and his eternal optimism, are gone. Science has been harnessed to human endeavour in the search for the earth's riches.

But the small prospector is still busy and, because of the massive amounts of capital required to bring a new orebody into production, he no longer works entirely on his own; when he finds something interesting he usually takes his discovery to one of the big companies which will have the financial resources to develop it, if it is proved to be viable. A survey made in the 1970s showed that in the first five years of the decade, an average 2,500 submissions were made each year to 40 major mining companies and, of these, 85 per cent were from small miners. Nearly half of these submissions were further investigated by the companies concerned, with the result that an average of 15 mines were being planned or brought into production annually as a result of the discoveries of the small miners.

With the growth of American mining, a number of newspapers and magazines sprang up, some of them specifically to cater for the needs of the small operator, others to cover mining interests on a regional or State basis, and some to do both. In such circumstances were born periodicals like *Skillings' Mining Review* of Duluth, Minnesota; *California Mining Journal; The Mining Record* of Denver, Colorado; the *Arizona Mining Journal* and *Pay Dirt*, of Bisbee, Arizona; and *American Gold News* of Ione, California. There were many others. Some did not last for very long, collapsing for lack of capital or becoming merged in larger and more nationally-orientated papers; but a few, like *The Mining Record*, now in its 90th year, and *American Gold News*, which has been going since 1933, are still flourishing, together with a clutch of newcomers to the field.

At the time of writing, these papers were united in one of the most vigorous campaigns in their history—a fight against proposals before Congress to repeal the 1872 Mining Law, which permits any United States citizen to explore for minerals in Federal land and locate a claim on his discovery, and to replace it with a system of competitive bidding for leases. The small miners' Press urged readers to lobby their Senators and Congressmen against the proposals which, it was claimed, would put the small man out of business, or at least relegate him to areas of little mineral potential.

In *The Mining Record*, J.K. Richardson, president of the Arizona Mining Association, declared: "There are some of us who continue to find merit in old values, such as morality, experience, dependability, integrity and courage! As a young man, I recall the revolutionary turmoil of Russia during, and immediately after, World War One. Emerging leaders constantly worked to eliminate the independent small landowners and place control of all lands in the hands of a bureaucratic system—and they succeeded. If you control the land, you control the resources; and control of the resources results in the control of the nation. The current mining law, providing the private citizen access to, and tenure upon, public lands, stands in the way of forces at work within our own nation to nationalize all industry. If these forces prevail, the day will come to states such as Arizona, where only 17 per cent of its land is in private hands, when government will be the sole source of all natural resources as more and more lands are withdrawn at bureaucratic whim."

Arizona has always been noted for its straight and forceful talking, and among its older mining men there is still affectionate remembrance of its *Mining Journal*, which was such a steadfast champion of their causes between the two world wars, and whose work has since been faithfully carried on by *Pay Dirt*, which is imbued with the same spirit.

The story of the *Arizona Mining Journal* and *Pay Dirt* is one of the many romances of that brand of mining journalism which operates among the grass roots of the industry. It is a story full of interesting and unusual characters, including that of a remarkable woman who played an outstanding part in the development of both papers; to do justice to its telling would require a book of its own.

The Journal cannot be said to have had an auspicious start. Its first number was dated June 1917 but, after two issues, its two enterprising promoters vanished for destinations unknown, taking with them the revenue from all advertising accounts, but leaving unsettled their account with the printer, Mr. R.A. Watkins. Despite his disenchantment with the original promoters, Mr. Watkins was convinced that the paper had real possibilities of success and, since he already had a sizeable stake in the venture in the form of his unpaid bill, he decided to continue publication. Under his management—and with the continued use of his printing presses—the magazine was produced monthly from August 1917 through December 1919.

During these months one of its most distinguished and regular contributors was Charles F. Willis, professor of mining engineering at the University of Arizona, Tucson. Willis had come to Arizona in 1912 when the state was newly-born and conditions were rough and raw, in direct contrast to the mellowed and cultured atmosphere of Boston, where he grew up. He graduated at the Massachusetts Institute of Technology, at nearby Cambridge, and attracted some notice as a musician: he once composed a pop musical score and directed its performance by the famed Boston Symphony Orchestra. Whilst at the University of Arizona, he established the Arizona Bureau of Mines and became its first director. It is obvious, too, that he soon recognized in Mr. Watkins's infant journal a real potential, both as a private business of his own, and as a public relations tool whereby a better understanding might be fostered between the local mining industry, then on the threshold of great development, and the general public.

Willis left the University in 1919 to become supervisor of industrial relations for Phelps Dodge Corporation at Bisbee, but a year later he resigned, bought the Journal from Mr. Watkins and took over as editor and publisher. Under his direction the paper prospered; publication was changed to fortnightly and editorial coverage and circulation were expanded to include most of the Western states—and, for a short period, Alaska, Mexico and even the Philippines. Within a few years, "Arizona" was dropped from its title (without seeking permission from its London namesake) and the paper became known simply as the *Mining Journal*, Phoenix; it continued as such until 1946, when Willis sold to Miller Freeman Publications of San Francisco, publishers of *Mining World*—with which the Journal was merged—and other magazines devoted to the natural resource industries of the Western states.

Among the Journal's many contributions to the mining industry of America's West was its sponsorship in 1938 of the Arizona Small Mine Operators Association (ASMOA). Willis was named its state secretary and continued to direct its activities for some 30 years until forced by illness to reduce his efforts. *Pay Dirt* started life as a bulletin for members and was soon adopted as the official ASMOA publication, acquiring the status of a registered newspaper. The paper quickly won wide readership, not only in Arizona, but throughout the United States: wherever non-ferrous metals mining people gathered, as well as in the halls of Congress and many state capitals, *Pay Dirt* was regarded with respect as an authority. Few people, however, realized that for much of its history, which has now spanned 40 years, the voice of its authority was that of a woman.

Although the by-line R.G. Moore was always used sparingly, whenever it appeared it was a guarantee that what was printed beneath would be an objective analysis of whatever matter was currently under discussion, and any opinions expressed would be with force and clarity. The paper's owner was a strong believer in editorial anonymity. One day in the mid-1930s a mining engineer remarked to Charles Willis that he would doubtless be surprised to learn that the associate editor of a magazine in San Francisco, for which he frequently wrote, was a woman. Willis replied that he was not a bit surprised, and added that for a number of years the editor of *Mining Journal* and his "right-hand man" had been a young woman. The mining engineer could not suppress his astonishment. "You have", he said, "destroyed one of my cherished illusions. I always believed the R.G. Moore of *Mining Journal* to be a man who smoked a smelly pipe and did his best thinking with his heels propped on the desk."

Rosalie Genevieve Moore—Genevieve to her friends—has continued to shatter illusions throughout most of her life. Many an aspiring reporter has lamented: "R.G. Moore is one tough editor." Many a glib publicist has remarked: "R.G. Moore is one of the hardest-nosed editors in the business." To Genevieve, now in retirement in Phoenix after 50 years of mining journalism, these have been the accolades that made an often-difficult job rewarding, and a lively sense of humour served over the years to maintain that perspective and detachment vital to successful editorship. Throughout it all she was supported by the rare gift of being at home in any company, yet remaining always herself. At American Mining Congress conventions and at other gatherings which often included some of the country's toughest and most practical-minded miners, she was always sought for her charm and wit, and for her knowledge and understanding of the industry.

Genevieve became associated with mining and publishing simultaneously and, like so many other successful editors, "by the back door, by accident". After graduation at DePauw University, she had a number of jobs, including a year as a teacher—a career traditional in her family—and in 1923 she applied to the Journal, where she had heard there was an opening on the editorial staff. Charles Willis turned her down because of her lack of experience, but he found her a temporary job with an organization with which he was associated. One day she heard Willis puzzling over an impossibly full schedule; he had meetings all day and news releases which had to be written and ready for mailing by evening. She volunteered to help and, in desperation, he agreed. It worked. Not long afterwards another vacancy occurred on the Journal—and this time Willis offered it to her.

The next few years were full of hard work in which Genevieve had to learn her craft. She worked, she says, "with *Webster's* at my right hand, a glossary at my left, the copy on which I was working in front of me, and the telephone within easy reach to call for help from anyone I could get it from. And, fortunately for me, I always found mining men and printers to be most gracious and generous in their willingness to supply help. Without them I don't know what I would have done."

She soon became associate editor and, as the years went by, she took more and more editorial control. When Willis sold the Journal to Miller Freeman, Genevieve was invited to go with it and work for *Mining World* in San Francisco. Alternatively, she could stay on as editor of *Pay Dirt*. It was a difficult choice, but Arizona was her home and she decided to stay.

Although she served *Mining World* as a correspondent, the demands of *Pay Dirt* took more and more of her time as it grew increasingly influential. Never a large circulation paper, it developed a disproportionately authoritative voice which was listened to by the decision-makers across the country. *Pay Dirt* had always placed special emphasis on the small-mining segment, but Arizona had now become the nation's largest producer of copper and a high-ranking producer of gold, silver and other metals; it was, moreover, the scene of operations of some of the country's most powerful companies and a centre of modern mining technology. It demanded a paper of high standards and integrity. Over the years, Charles Willis, a remarkably versatile man, devoted more and more of his seemingly boundless energies to a variety of other activities, leaving the paper almost entirely in Miss Moore's hands. Today she acknowledges that the pressures were often more than she bargained for; but she still appreciates the compliment.

In the field of mine safety, Miss Moore feels that both of the publications she edited made useful contributions. At one time the company had an arrangement with a miner in Montana, who was safety-conscious and had a gift for cartoon drawing, as well as a good sense of humour. His cartoons were published in the Journal, after which major producers were persuaded to finance their enlargement and reproduction on heavy gloss paper for posting on all bulletin boards at the mines. The humour and the cartoon presentation brought the safety message home to many who would have ignored a conventional caution. Letter-size folders were also produced with each page illustrating a hazard that might be encountered in entering and attempting to explore the old and abandoned shafts and tunnels with which the desert country abounds: rattlesnakes and scorpions, rotten boards and pillars, and hidden waterholes. These folders were distributed to school teachers and other youth group leaders who took their students on field trips and to summer camps. The folders were so popular that they were printed in batches of 5,000 and 10,000 and distributed from Arizona to Washington State. The same material was also used by Arthur J. Flagg in his treatise *Mineral Collecting* and in safety bulletins issued by the U.S. Bureau of Mines.

Charles Willis died in 1968 after a prolonged illness. During his last months, Miss Moore urged that *Pay Dirt* be sold to assure its continuation and also to enable her to retire and take some well-earned rest. Willis agreed, but only on condition that she find a purchaser who could be trusted to maintain the same standards that he and his editor had set. There were several offers, but none of them seemed to meet the essential requirement, and it was decided to suspend publication. An announcement to this effect was drafted and published—for Miss Moore it must have been like writing her own obituary—but at the eleventh hour came reprieve. Through some old ASMOA friends, two Arizona newsmen and publishers, William C. Epler, of Bisbee, and William Bideaux, of Tucson, were brought together with Willis and a deal was made. Epler later became sole owner and Miss Moore agreed to stay on to assist in the transfer of interest and to guide Epler in his relations with the smallworkers and contacts with the big producers.

The new arrangement has resulted in great expansion for *Pay Dirt*, both in size—it now runs to an average of 64 pages per issue—

and in scope, aided recently by the launching of a New Mexico edition.

Miss Moore eventually retired in 1973, at the age of 78, after completing her half-century in mining journalism because, as she said, she needed more time to devote herself to other affairs and interests! She is delighted with *Pay Dirt*'s performance in the 10 years since Bill Epler took it over. "The paper", she says, "fits the needs of the industry today much better than it would have as originally conceived. Bill has, without any sudden, drastic changes, converted it into something very like the *Mining Journal* used to be. It is a dynamic, accurate, informative paper which is now big enough to cover all of the many important developments in Arizona mining—and Arizona mining is so important to American mining that the publication is important reading for everyone in the country who must keep current with the industry."

Of the many tributes that have been paid to Genevieve Moore's work for mining and mining journalism, two are particularly noteworthy. The first came from the contemporary *Arizona Republic*, of Phoenix, when *Pay Dirt* came under its new ownership in 1968. "Under Miss Moore's direction," said an editorial, "*Pay Dirt* down through the years agitated for the preservation of a pioneer character in danger of being crushed between big government, big business and big labour. *Pay Dirt* championed the small miner. The new owners may make it bigger. But they will be hard put to make it better." And, when Genevieve eventually put down her pen, the Governor of Arizona, Jack Williams, wrote to her: "The initials R.G. are familiar to thousands of persons of two generations who are, or who have been, involved in mining activity, and they are going to be missed. Your career has been good for Arizona; it has benefited one of our major industries and consequently our entire economy. You are one of the builders."

CHAPTER SIXTEEN

The Media and the Market – I

One day in 1876, a group of United States jobbers, meeting at one of those conventions which were already becoming much loved by American businessmen, expressed deep concern that the publication of hardware prices in the magazine *The Iron Age* was having an adverse effect on their trading. One angry member declared: "These price quotations are simply ruining our chances of doing business. The paper goes to every crossroads hardware store in the country and gives the dealer advance notice of every decline in prices. We cannot do business in the face of competition like that." The convention felt so strongly about the matter that its members resolved that they would no longer handle the products of any manufacturer who advertised in the paper.

The Iron Age not only printed a report of the convention, but published the names of all those who had supported the resolution to boycott its advertisers. It went even further. Its editor turned up and published a report of another resolution, passed by hardware jobbers 20 years earlier, pledging that they would refuse to handle the products of any manufacturer who marked his goods with his name or trademark. "Where are these jobbers now?" asked *The Iron Age*. Only two of those who were trading in 1856 appeared still to be in business. It was a sharp reminder of the boomerang effect that such boycotts, emotionally imposed, can have, and, in fact, the threats of the 1876 convention did the paper far more good than harm.

In its price reporting, *The Iron Age* was continuing a policy which was deep-rooted in its foundation in 1855 by John Williams, an Irish hardwareman whose revolutionary political activities made it "expedient" for him to leave his native land. He emigrated to the United States and from a little print shop in Middletown, New York State, he soon began publishing *The Hardwareman's Newspaper*, which in a few years was to become *The Iron Age*, based in New York City. His paper was probably the first anywhere in the world to make an important regular feature of price reporting, and its success undoubtedly inspired the development in later years of other periodicals, on both sides of the Atlantic, which were to be specifically concerned with the reporting and, in many cases, the evaluation, of metal prices. And out of such development there grew an international service which has become one of the most significant contributions which journalism has made to the metals industry.

As a pioneer in this field, and for its chronicle of the United States ferrous metals industry from its formative years, *The Iron Age* has earned a high place in industrial journalism. Whether its foundation was a stroke of vision or a stroke of fortune, it is difficult to say; perhaps it was a touch of both. The time was certainly propitious and the environment favourable for growth. Five years earlier, the value of the country's manufactured goods had for the first time exceeded that of its agricultural products. In 1855, the United States was on the crest of a great wave of expansion, due largely to the gold in California and the general westward movement. Railways were becoming the trunk routes to link the two great oceans. A year after John Williams started publishing, the tracks had reached Kansas City, nearly half way across the continent, and on the Pennsylvania railroad a white-haired little Scotsman, named Andrew Carnegie, was working as a telegrapher for divisional superintendent Tom Scott. Later, Scott became president of the railroad and his nervy little telegrapher drifted into the steel business.

The Iron Age grew, as the industry to which it was linked grew, by leaps and bounds. Despite the opposition of the jobbers, John Williams clung to his resolution to publish prices of hardware items in an effort to break down the secrecy under which most deals and concessions were then made, and to create a fair market for all. His son David joined him in the enterprise and when his father died, it blossomed under his direction. In the early days, the hardware trade had been the major metalworking field, but the iron and steel industry was a robust infant, and by the turn of the century the paper had expanded its market reporting to embrace all ferrous products, including ore and scrap.

When it celebrated its 40th birthday its well-wishers included the former Pennsylvania railroad telegrapher who now controlled the country's biggest steel plant, fed by tributary coal, iron fields, railroads and a fleet of lake steamers. "For 30 of those 40 years", wrote Andrew Carnegie, "I have been one of the pilgrims on the march with *The Iron Age*. A checkered and crooked path, indeed it has been, over bogs and morasses, and through floods which swept many comrades away; nevertheless, the vast host has moved steadily in the right direction. In learning by experience much-needed lessons, by availing itself of every scientific discovery and every new source of supply, and adapting itself to the wants of the new Republic, the undisciplined few who began their wanderings under your guidance have become the largest, best disciplined, and most effective army of ironmasters in the world ..." Six years later Carnegie's companies were incorporated in the United States Steel Corporation and he retired from business to give effect to the doctrine which he expounded in his famous essay, "The Growth of Wealth", distributing his vast personal fortune for the advancement of education, social welfare and international peace.

When David Williams sold *The Iron Age* to a group of business paper publishers in 1909 he had controlled its affairs for more than 50 years. It was thought by those who had a glimpse of his private papers that he wrote his records in Arabic to keep them safe from prying eyes, and, because he had been so successful, there were many whose curiosity ran high. But it was not Arabic which ensured confidentiality; it was Williams's own version of Mr. Pitman's shorthand.

It was a sign of the times that the new owners made a separate magazine out of the hardware coverage and placed heavy emphasis in the parent paper on metallurgy and on new machinery, inventions and plants. By the 1930s this policy had been extended to include seeking out technical data which some people preferred not to see published, but which the paper considered valuable to the majority. Subjects which were thus explored and tenaciously followed included continuous casting, electrolytic tinplate, stainless steel, the flame-hardening process and many new types of rolling mills. One of the first of these "scoops" was a 1936 report on electrolytic tinplate—"the first radical development in tinning since the year 77 AD". Details were highly secret and there was much speculation about companies and processes. As it turned out, electrolytic tinplate came just in time to save the United States from a serious food shortage in the Second World War—and *The Iron Age* gave a complete report on the process which so drastically reduced the need for the tin supply which Japan had cut off.

Except for the work done between 1846 and 1857 by Sir Henry Bessemer, the first published report on continuous casting appeared in *The Iron Age* in 1935. Until then this process of going from the liquid metal to the rolled product had been in secret commercial use for several years in the production of certain non-ferrous metals. This first report had a significant effect on the industry. Readers also received the first report on United States developments in cold extrusion of steel. Boron steel developments were another area where the paper took the lead in bringing specific metallurgical information and case histories to those who urgently needed such data during two war emergencies.

Contrary to popular belief, a relatively small proportion of the magazine's technical pages have been devoted to steelmaking over the years. And a much larger proportion of space than is generally imagined has been given to non-ferrous metals. For example, titanium received extensive coverage in the 1950s when it was one of the newcomers in this field, and more recently subjects such as ductile iron, powder metallurgy and vacuum melting have been followed. The paper's editors have always felt that although production figures or invested capital may be small in some processes, it is only by reporting developments in infant industries that their potential can be fully examined.

This habit of digging new ground for new ideas has gone on continuously since the paper started. Early in the present century, information on pig-iron production for both foundry and steelmaking use was hard to obtain locally—and impossible to get nationally. In 1901, *The Iron Age* began gathering its own monthly statistics. It took much time to overcome the reluctance of blast furnace operators to divulge their production figures confidentially to the paper so that an overall figure could be worked out without revealing the production of specific companies. But eventually their confidence and their co-operation were won.

The monthly pig-iron figure came out in the first issue of each month and the editorial staff frequently worked all night compiling totals from information wired in from all parts of the country. This sort of "newspaper" speed, commonplace today, was most unusual in a "trade" magazine at that time. But it was worth the effort, for these statistics were used by salesmen, business forecasters, ore companies, steel firms and steel customers in helping to determine market conditions. Moreover, the pig-iron figures became one of the first reliable barometers of business conditions, not only in the iron and steel market, but in domestic industry generally, and for many years it remained an important factor in assessing the state of the economy. Production figures were divided into 12 geographical areas, thereby reflecting the business temperature in each part of the country. When blast furnaces shut down or started up, it meant far more than the movement up or down of the consumption of coke, ore and limestone; it pointed the way to the ups and downs of the corner grocery shop and the high street bank. Understandably, during the Second World War, government authorities decided that such information could give "aid and comfort" to the enemy and it was accordingly blacked out. At the end of the war, the American Iron and Steel Institute took over the compilation of pig-iron statistics, and although its methods of reporting were not the same as previously used by *The Iron Age*, the continued publication of the figures was a tribute to the magazine's work in this field—and confirmation of their value.

In years gone by many hundreds of thousands of tons of pig-iron and finished products were bought and sold in the United States on the basis of *The Iron Age* quotations and, even after the major steel companies began to publish their own prices, the paper's quotations continued to be used in countless numbers of deals involving materials far removed from ordinary steel products.

Inevitably, not all the magazine's price quotations have been unanimously accepted, and at times—particularly in periods of market depression—published figures have been vigorously challenged. Not surprisingly, this has been particularly so in the scrap market, where the process of arriving at a "going price" for the various grades is a most exacting one. But the large number of contracts made at the end of the day on the basis of the published quotations reflected the acceptability of the figures.

The Iron Age gave up market reporting some 10 years ago, but still reports prices. If the paper has at times had a somewhat "heavy" reputation, it is not because its technical articles are any weightier than those which can be easily digested by readers who have the stomach for this type of literature; rather has it been on account of the size and thickness of its issues. As long ago as the 1870s, it was a joke among subscribers that if they wished to go on having the magazine they would have to move into bigger offices—and this was a dilemma which was also to face *Mining Journal* readers in Britain some 30 years later. In New York, the paper's reputation spread far beyond those concerned with the metals industry. In the 1920s, Broadway theatre-goers were amused by a scene in a play in which the script called for one of the characters to be hit on the head with something exceptionally large and heavy. The cry was: "Hit him with a copy of *The Iron Age*!"
It nearly brought the house down.

*

At about the time that the American hardware jobbers were attacking *The Iron Age* for its audacity in printing prices, a small group of metal traders in London was negotiating the hire of a room in Lombard Court for the use of the newly-formed London Metal Exchange Company. By the 1870s, London had become the centre of a vigorous and well-established market in base metals, and in its transactions a distinction was already emerging between a physical market, where metal was bought and sold for delivery, and a market in "arrivals" where hedging and speculating were complementary activities.

The founders of the LME recognized that one of the main benefits of the trade meeting together was the establishment of common prices; hitherto the reporting of prices had been totally unorganized; indeed, there had been little pattern in the history of metal marketing in the 5,000 years since the simple bartering of primitive tribes who first discovered the secrets of mining and smelting, and the uses to which the finished material could be put. Today, the LME and, to a lesser extent, the New York Commodity Exchange (Comex), provide highly sophisticated terminal markets for some of the world's major metals. For example, most of the world's supplies of newly-mined copper are sold either at prices which are fixed independently by producers, or at prices which are related to quotations on one of the commodity exchanges. For all practical purposes, the latter meant, until recently, the LME; at the time of writing, however, the whole marketing position has become confused with the crumbling of the producer price system in the United States following the adoption of Comex as a reference price by some of the major groups. In addition, renewed attempts are being made by the governments of copper mining countries in the developing world to evolve a new formula aimed at bringing some stability into what has historically been a highly fluctuating market—naturally, the aim is for "stability" at a considerably higher price level than that which the free market has been reflecting over the past few years.

As with the marketing of any other commodity, the law of supply and demand must be at the root of all metal trading. By its very constitution and operation, the LME should reflect the balance of this situation at any one time but, of course, it can do so only in relationship to the tonnage which is actually sold on the Exchange. In the case of copper, this is a very small proportion of world primary production, and consequently critics of the Exchange maintain that it can never be a true barometer of world price. In contrast to the terminal market, the producer price system—and again the reference here is specifically to copper—tends to reflect a price which is not strictly determined by supply and demand, but is more an assessment of the balance of production and consumption over a period. The two situations are vastly different, though in "normal" conditions, the price produced by the one system is very often roughly in line with that shown up by the other. But the copper industry very seldom enjoys "normal" conditions and, indeed, it would be very difficult to define the word "normal" in this context.

Logically, the full production of any one metal should sell at roughly the same price, no matter how it is marketed—and, by and large, if there are not too many "abnormal" factors, this is what happens in practice. But the costs of production of those who supply the market vary enormously, and the general pattern in the copper industry is one of low-cost producers making a lot of money for their shareholders, and high-cost producers just getting by when the price is high, and facing disaster when it slumps. Mining is not the sort of operation which can be turned on and off like a tap: a company whose mine becomes flooded when development is suddenly brought to a halt may have to wait 20 years before the economic climate enables it to get back into business; if the company is producing a strategic material, it may have to suspend operations until wars or threats of wars put the product at a premium again.

There are, of course, very many commercial metals in use today other than those which are either dealt with on the terminal markets or are producer-priced, and dealers in such metals need some reliable reference to help determine prices. It is in this field that the specialized Press today plays a particularly important part, its systems of reporting and evaluation helping to establish current prices for scores of lesser-known metals, alloys and semi-fabricated products. Contracts for the sale of considerable quantities of the major metals also continue to be written on the basis of quotations published by the various papers when sales are negotiated outside commodity market, or producer-price systems.

To *The Iron Age* must go the credit for taking the first positive steps in the United States to bring some order to the hardware market, which had so often been cloaked in secrecy and distrust, and later in helping to build up a service for those engaged in the wider business of buying and selling metals. Other papers were not far behind.

From its very first issue in 1866, the *Engineering and Mining Journal* published market reviews and price lists from the New York and London markets. Elaborate statistical issues in January, 1891 and 1892, were followed by Richard Rothwell's monumental work, *The Mineral Industry—Its Statistics, Technology and Trade*, which was launched in 1893 and published annually for the next 50 years (see Chapter Five). Walter Ingalls, a statistician and protégé of Rothwell, who assumed the editorial chair after the four chaotic years which followed Rothwell's death, strengthened the statistical department and established what came to be known as the paper's Metals Price Quotation Committee, consisting of two *E/MJ* market editors and a representative of the Department of Commerce. Ingalls gained the confidence of producers and personally canvassed the New York market every week in order to obtain current prices of a big variety of metals. In Ingalls's day, a large proportion of all metal sales contracts in the Western World were made on the figures carried by *E/MJ*. The paper's authority in this respect was recognized not only in the market, but accepted in law: in a famous court case just after the First World War the judge observed that "The *Engineering and Mining Journal* was, and is, recognized as a standard and reliable source of information . . . for the daily current market price of metals."

Josiah Spurr, who succeeded Ingalls in 1919, took things a stage further. He and his colleagues became deeply conscious of the fact that of the four main divisions of the mining industry—geology, mining engineering, metallurgical engineering, and marketing—marketing had been virtually neglected from the point of view of scientific, or even planned, development. They resolved to wake up the industry to the need to market its products properly and to apply the same sort of discipline to these operations as in the production processes. With the co-operation of leading experts, a series of more than 60 articles on the marketing of 92 metals and minerals was published in the magazine. These articles were later put together in a book, first published in 1925. Mining schools, stirred into providing students with marketing instruction as part of the curriculum, used this book as a text, and it became a standard work on the subject.

By this time competition in metal market journalism was becoming quite lively. The *American Metal Market* had been formed as a weekly news and price paper as long ago as 1882. A year or two later it was bought by Charles S. Trench, a New York metals broker, who turned it into a daily and, according to available records, it has never since missed a scheduled publication date; it remains today the world's only daily newspaper in this field. Other market papers sprang up in the United States, among them the *Daily Metal Reporter*, founded just before the First World War.

Outside North America, the London-based *Metal Bulletin* was another war baby, though the seeds of its production had been sown a couple of years earlier by L.H. Quin, an Irishman like John Williams who founded *The Iron Age* 60 years earlier, and who had also been associated with the hardware trade—not like Williams, as a businessman, but as a journalist on the staff of *The Ironmonger*.

This mounting competition was obviously an influence in *E/MJ*'s decision to launch a separate publication to cover price quotations and other market information. The weekly *E/MJ Metal and Mineral Markets* made its first appearance in January 1930, produced by a small staff under the direct supervision of the editor of the parent magazine, then Arthur W. Allen. The new paper aimed to supply price quotations and informed market information for producers and consumers, and among buyers and sellers of raw, intermediate, refined and scrap material. It was mailed to subscribers after the closing of the market every Wednesday, its production being untrammelled by the delay inherent in the printing and binding of a technical magazine. Introduced as a service to industry, it represented a new form of communication—one which has developed into the modern newsletter, and, in some applications, into a glut of highly-expensive "inside" or "confidential" services, many of which contain little more information than is carried by the daily Press.

Basically, the service inaugurated by *E/MJ* in 1930 has continued ever since, even though the vehicle has undergone a number of changes. In course of time *Metal and Mineral Markets* outgrew its early format; in 1967 it changed its name to *Metals Week* and in 1972 it was pruned to its original format, shed its advertising and became a newsletter again.

Throughout most of its history, the method used by *Metal and Mineral Markets* in arriving at its price quotations was little more than an extension of that introduced by Ingalls more than 10 years earlier. Most recognized producers and many consumers of major non-ferrous metals reported their tonnage and price details covering sales for each day of the previous week. The market editor then calculated a weighted average price for each metal on a daily basis. Weekly and monthly average prices were also worked out arithmetically to assist buyers and sellers with long or short-term sales contracts to settle at prices which accurately reflected the market. The market editor also obtained sales reports from foreign producers of certain non-ferrous metals, from which weighted averages and daily "export" prices were prepared. *E/MJ* made much of its claim that such sales statistics were not reported to any other publication or agency, that they were highly confidential figures given exclusively by producers and sellers because they knew from long experience that the paper would respect their confidence.

There is no doubt that at peak periods hundreds of millions of dollars worth of refined metals, ores and scrap were being bought and sold all over the world in deals which specified *E/MJ* prices as a basis of settlement. In many cases these quotations have also been an important factor in determining wage scales or bonus payments in mining companies in which employee-contracts have had some relationship to metal prices. The United States Government has used *E/MJ* quotations to fix price levels in connection with tariff laws, and overseas governments have made use of its figures as a basis for levelling export taxes on metal shipped out of their countries.

Alvin Knoerr, who edited *E/MJ* from 1955 to 1969 and had direct editorial control over *Metal and Mineral Markets*, once said: "We get nothing out of it except the responsibility of doing an accurate job, and that's a responsibility we are very proud to have." Accuracy had always to be the keynote because a mistake, say of half a cent a pound in copper, would mean that someone was getting several thousand dollars more—or less—than he was really entitled to. Personal integrity of the market editor and his assistants was another factor. "They simply have to have the degree of integrity that cannot be questioned," declared Knoerr, "and we are fortunate in getting that kind of people. It is made very clear to a man before he comes on the job that if he cannot live up to that trust it is simply no use. The job is not rewarding in any financial way . . . integrity doesn't carry a price . . . the fellows who do the market work don't get a tremendous salary, but they are probably the most influential people in determining the price for a pound of copper.

"Every once in a while something will come up that has never happened before, and we may have to exercise judgement, and here again we have only one target, and that is to be absolutely right. There's nobody to whom we can appeal, not even the Supreme Court. Once we had a situation where it became apparent that Chile was demanding a price of 50 cents a pound for copper when she wasn't selling any—and the true market price was around 35 cents. In other words, Chile was attempting to force a price that was unrealistic, and it was only because of an extremely short situation that she was able to try this. Our market editor at the time, and our chief editor, Bob Ramsey, had to make a decision whether to include the Chilean price in the current calculation or leave it out. They knew that their decision could create an international situation. They paced the floor right up to the time they had to let the paper go to press. They decided to throw it out. They could have been wrong, and they could have been right. There was no one else to tell them." Incidentally, it did happen that they were right.

E/MJ's self-imposed obligation to serve the industry in this way inevitably produced a host of problems. Quotations were often hotly challenged and the paper had to appeal to producers for documentary evidence of reported sales—again in confidence, of course—in order to substantiate their prices. This was invariably given willingly and promptly.

Today, for reasons that have already been explained, there is not the same call for quotations of the major metals as there was 20 years ago, and some of the published figures tend to be rather academic, but *Metals Week* carries on the tradition of its predecessor with "insider reports", a telexed price notification service; and weekly quotations for some two dozen minor and light metals, precious metals, and a string of ferro-alloys, as well as the old stalwarts, copper, tin, lead, zinc and aluminium.

CHAPTER SEVENTEEN

The Media and the Market — II

The Ring at the London Metal Exchange

American Metal Market is a newspaper. It *looks* like a newspaper; it *feels* like a newspaper; and, to anyone who has grown up in the newspaper business, it *smells* like a newspaper; even when it tumbles into the hands of its readers after going through all the processes of its creation, it still carries a whiff of that exotic aroma of newsprint and ink that would enable a blind man to recognize it for what it is.

To some extent, the same might be said of Canada's *Northern Miner*, but *American Metal Market* is published every day of the American business week, and the urgency and the pressures attending its production are therefore much greater than are exerted upon a weekly—though not five times greater, as some would claim. It has been a daily paper for 80 years and its record of continuous production is unique in the world of metals—and in most, if not all, other international industries, too. Its policy is a very simple one: to bring the hard news of the metals business to those who need it as quickly as possible. It is not a campaigner; it has no axes to grind; it does not even carry an editorial or opinion column. It is, in fact, one of the few papers left in the Western World which concentrates so strongly on news—radio and television having largely taken over this function, leaving the printed Press to fill in the background and offer comment and commentary.

And, whilst in these days of "unisex" and "women's lib" one must be careful not to discriminate, it must be said that its present editor, who is considered by many to be one of the most accomplished

209

journalists ever to be associated with the paper, just happens to be a woman. Patricia L. Walker (Pat, as she is known by her associates) is probably the first woman to be top-editor of a daily business or industrial paper that is not orientated to a largely feminine readership.

American Metal Market was born out of the activities of William I. Russell and Company, who were one of the leading New York metal brokers of the 1880s, specializing in pig tin. Russell is believed to have made something of a fortune in the hey-day of the Secrétan Syndicate which was set up, with the backing of French bankers, to corner world supplies of copper and force up the price—for the benefit, of course, of those members who had bought before the rise. When the bubble burst the supporting banks collapsed, the price plummeted and the syndicate went to the wall.

William Russell had many ups and downs, principally downs, and it was in one of these low periods that he sold his paper to Charles S. Trench, whose company was also in the brokerage business in a big way. The paper was then a weekly, featuring prices and very little else; from sheer neglect it had slumped from early success and considerable popularity to near-failure and obscurity. For a year or two, under its new ownership, it made little headway and it is doubtful if its limited revenue was enough to cover the bare expenses of printing and distribution. In 1899, Trench decided to publish daily in direct competition with a number of other market reports, notably the official report of the New York Metal Exchange and the *Daily Market Report*, produced by James E. Pope, another member of the brokerage fraternity, under the title of The Report Company of New York City.

Jim Pope appears from the records to have been a shrewd business operator of striking looks and character. He has been described as a tall, piercing-eyed, moustachioed man who favoured immaculate celluloid collars, and who lived to enjoy the distinction in the 1920s of being the oldest alumnus of Yale University. He saw the possibilities of his time and made the most of them. Each of the two hats that he wore, of broker and of publisher, fitted him neatly, but in 1901 he discarded the latter and sold out his 11 year-old paper to Charles Trench. Trench incorporated his acquisition with *American Metal Market*, doubled its size, and upped the subscription from $7.50 to

$10 a year. As part of the deal, he also took over one of Pope's employees, 21 year-old Fred Behrje, who was to remain on the staff of the paper for another 50 years.

At the time of this first take-over, the size of the composing room in which *American Metal Market* was produced was 15 ft. × 20 ft., and the editorial office was no larger than a proof-readers' room. One of the issues in 1901 invited subscribers to visit the company when they were in the neighbourhood; it is hoped that they did not have too many acceptances as four people, busily at work at three desks, completely filled the office. Trench's own company supplied the tin report and attended to what little book-keeping was necessary—and Jim Pope continued his association by providing the copper report.

Very little editorial material that went into the paper at that time was paid for. C.S.J. (Jack) Trench, one of Charles's three sons, all of whom became active in both of the family businesses of brokers and publishers, confessed years later: "We would crib unmercifully from *The Iron Age, Connellsville Courier, Iron and Coal Trades Review*, and the Dun and Bradstreets report: and, for a large amount of the material we printed on non-ferrous metals, we depended on the circulars issued by metal concerns both in New York and in London. We also regularly and religiously reprinted the Rogers Brown and Mathew Addy pig-iron reports, regardless of whether they contained anything of real interest."

Early progress was certainly not spectacular and it was not until 1906 that subscriptions topped the 2,000 mark, but the foundations were solid and the policy was sound. From the outset, the paper concentrated on simplicity. The overriding object was to gather the latest and most accurate price information and publish it with the utmost clarity: neat rows of figures commended themselves to the narrowed eyes of broker and merchant; quotations were as brief as the typographer's art could make them; words were always expendable when figures could tell the story. Even advertisements followed the same succinct pattern: "Burge Brothers—Cliff Street—Zinc".

Charles Trench was an Englishman—"English to the nth degree", according to one colleague, and he remained a British subject until his death in 1926. Thereafter, until the early 50s, his son Jack became the mainstay of both businesses, with the help of his brothers, Archer R.R. and Stewart P. Trench. Archer was principally concerned with the paper until he joined the Canadian Army in the 1914–18 War, during which he died. Stewart returned from service abroad with the British Army and was in and out of the business until his death in 1953.

Jack Trench's sons, Charles S. and J. Patrick Trench, both entered the business briefly before the Second World War and returned to *American Metal Market* when it ended. The brokerage business was not active during the war and it was never revived. In 1956, Archer W.P. Trench, son of Archer R.R., became chief executive of the company. By this time, in addition to the daily paper, the company was running a highly-successful price information service, by which subscribers were telephoned twice daily with LME and New York market reports, and it also continued to publish *Metal Statistics*, an annual review, first started in 1907. Even after the cessation of the family brokerage business, the paper maintained a staff of "experts" who covered only their own beat: Norbert Langer, copper; Jimmy Stafford, lead and zinc; George Ehrnstrom, steel (Pittsburgh); John Ruth, rare and precious metals; Pat Trench, tin; Karl Rannells, aluminium. Frank Smith, who was widely recognized as an authority on gold and monetary affairs, was editor from the 50s until the early 60s.

Competition was lively and in one period there were two other daily metal papers, as well as the weekly *E/MJ Metal and Mineral Markets* in the American field. The *Daily Metal Reporter*, one of Charles H. Lipsett's Atlas Publishing Company papers*, had been in existence since 1911 but did not make much headway until Dr. Joseph Zimmerman became its editor eight years later and set out to gain the confidence of top executives in the industry. For years he battled on alone, doing every conceivable job necessary to produce the paper, including selling a big proportion of its advertising space. Later, after he had become editor-in-chief of all the Atlas publications, he had the support of a small staff that included Si Wakesberg, now with the

---

*Charles Lipsett died in November 1978, aged 95. He published his first trade paper in 1905 and helped establish the U.S. Institute of Scrap Iron and Steel in 1928.

National Association of Recycling Industries, and Bill Hoffman, who succeeded Norbert Langer at *American Metal Market* and was its copper editor until he retired a couple of years ago.

There was never any suggestion of a David and Goliath battle between the one-man *Daily Metal Reporter* and the well-staffed *American Metal Market*, but the *Reporter* went a long way towards narrowing the gap, simply and solely because of the tenacity of Joe Zimmerman and the very high respect and personal regard which was accorded him by leaders of the industry, including the big producers, not only in the United States, but internationally.

"The *Reporter* did precisely the same thing as *American Metal Market*", Dr. Zimmerman reminisced some years after he had left editing to work more directly for the industry which he had already served so long. "We had a much smaller circulation, perhaps 2,500, compared with *American Metal Market*'s 12,000. Furthermore, their paper was ABC-audited, which was very grand indeed way back then. I am afraid that Mr. Charles Trench always considered the *Reporter* something of an upstart."

Like the rest of the industry, *American Metal Market* had great respect for Joe Zimmerman, even though they may not have feared that he would make any significant inroads into their circulation figures. Dr. Zimmerman knew that as a supplier of the same product of an old and well-established firm, he had to give something more for his subscriber's dollar. He offered two premiums:

"First, we appealed to the very common reaction against letting any one person or institution corner the market on anything. Then we stressed the personal service aspect. I myself went out into the field three or more days a week, and this man-to-man approach enabled me to assess a trend in the development stage."

Other than this, the main difference between the two papers was that whilst the *American Metal Market* was traditionally sensitive to the brokers and their activities, the *Reporter* became more and more responsive to the primary non-ferrous metals producers. This difference was brought sharply into focus when the big American copper producers decided to by-pass brokers, dealers and merchants entirely and deal directly with consumers, fixing their own prices. The producers declared that the main object of their decision was to stabilize prices, but in the brokers' view this was discrimination, pure and simple. At any rate, it provided much lively discussion and editorial controversy, and wrote another chapter in the chequered history of copper marketing. The other daily in American circulation between the Wars was the *Daily Metal Trade*, produced in Cleveland by Penton Publishing Corporation, but it was never a serious challenger and covered rather a different field, placing more emphasis on steel and ferrous metals generally.

The editorial objective of *American Metal Market* was always to let the news and the facts tell the story without comment by the contributors; the daily market commentary by the various market editors analysed the news, but did not take a position in relation to any issue. There was an editorial in which views were expressed—often pungently, especially in the days of Jack Trench and Frank Smith. Smith had a profound, though rather verbose, style and it was said of him that he sometimes sounded like a public relations man for the Democratic Party. There may well have been some substance in this remark as, earlier in his career, he had been secretary to Josephus Daniels, who brought the U.S. Navy to a high pitch of efficiency in the First World War. Daniels himself started life as a journalist and publisher before he went into politics; when he was appointed Navy Secretary he named Franklin D. Roosevelt as his assistant—and Frank Smith's office was between those of the two great men.

In the period between the two World Wars one of the paper's personalities was Colonel Percy E. Barbour, a mining engineer who wrote many editorials. The staff would quietly refer to him as the Sheriff of Esmeralda County, a post he had held in Nevada in his younger days. He was also secretary of the elite Mining and Metallurgical Society of America (no association with AIME). One of his outstanding editorials proclaimed that Hitler would lose the war because of Germany's shortages of copper and petroleum. He was partly right—but there *were* other factors.

The paper also had a "mystery correspondent". Practically nothing was known about him, except that he lived in Birmingham, England, and probably worked for a big metals consumer or a trading company.

John Ruth, who is today *American Metal Market*'s longest serving editor, recalls: "No one that we knew ever saw him and it was a mystery how he ever made contact with us. He always asked that his name should never be revealed. But his weekly report from Britain was the most religiously-read feature in the paper. Subscribers would call at the office if we were running late and they read his column as if they were addicted. He had a smooth, but lively, style . . . he knew his subject thoroughly and suggested possible trends. When he was near death he wrote us that his son was too stupid to carry on. Even after his death we never knew where he worked or what he did."

The past 20 years have been ones of growth and change in *American Metal Market*. Archer Trench's appointment as president-publisher brought to the company a combination of the "Harvard touch" and the bustling enterprise of an out-going activist. Archer had been concerned with the steel and metals industries since his graduation from university in 1937 and from Harvard Business School 10 years later. His career was interrupted by World War Two in which he rose to the rank of Commander in the U.S. Naval Reserve and commanded one of the famous British "Flower class" corvettes—*HMS Veronica*—on escort in the North and South Atlantic, and later a high-speed destroyer-transport in the Pacific. He was awarded the Bronze Star in the Okinawa campaign, and in 1969 he received from the Japanese government the Fourth Class Order of the Rising Sun for rescuing Japanese civilians at sea off Korea. Patrolling in these waters soon after the end of hostilities, his ship picked up 40 survivors, including women and children, from a floundering fishing junk, and brought them safely to shore.

When he was presented with his award in Tokyo, Trench took the opportunity of inaugurating one of the many innovations in *American Metal Market*'s activities for which he was responsible. This was a "briefing breakfast", arranged for American top-executives attending a meeting of the International Iron and Steel Institute. The American ambassador and three senior American business men in Japan talked to the guests about methods and customs of doing business in the country. The venture was a great success and was repeated in subsequent years in Toronto and in Paris, though a suggestion that a similar breakfast be held before the IISI meeting in London was vetoed by the British steelmen, presumably on the grounds that everyone knew, or should know, how the British did business.

A few years earlier, *American Metal Market* had begun its now well-known Forums with the object of bringing together those concerned with the production, processing and marketing of metals for a free exchange of opinion and ideas. These have now spread outside the United States to London, where, since 1969, the event has taken place each year during the week of the London Metal Exchange dinner, which itself attracts upwards of 2,000 of the industry's personalities, many of whom make a point of going also to the Forum.

"We laid down a strict formula for these Forums," recalls Trench. "The speakers had to be top-line executives with authority and first-hand knowledge, each from a different segment of business or technology of the metal under discussion: producer, distributor, consumer and broker. We spiced the programme with security analysts, economists and overseas guests. We never had a speaker who asked to speak. We sometimes ran into problems, as with the German who was to speak on copper and opened his address with 'I shall not speak on copper, I shall speak about scraps ...' We always tried to get one 'star' who would do his homework and attract a full house. Our biggest problem was with foreigners; electronic public address systems amplify the accent, but generally they ignored the microphone and mumbled through an unintelligible reading of their script."

Archer Trench probably probed deeper than any of his contemporaries, seeking always the "news behind the news" in the metals world. One of the ways he did this was with a series of "Newsmaker" luncheons—private interview sessions where senior executives were asked their views on subjects of particular interest to metal buyers. Leaders of some of the country's major companies joined the paper's editors in these sessions which produced a high level of informed opinion about current industry affairs and problems.

But the newspaper continued to be the backbone of the company—and, of course, the big money-spinner. In 1961, the company bought, and absorbed, the *Daily Metal Reporter*, which had quickly collapsed after Joe Zimmerman resigned to take up an appointment in the industry. With it, came the *Reporter*'s price service and the directories, *Mines Register* and *Standard Metalbuyers' Guide*—and Bill Hoffman, who became *American Metal Market*'s copper editor. The same year the company launched *Metal/Center News*, a monthly magazine for steel service centres: it did well from its first issue and was soon running its own seminars at home and abroad—notably in London, Tokyo and Copenhagen. In 1969, *The Oil Daily*, a national petroleum paper, joined the stable.

Two years later, the company sold its daily metals paper, together with the *Metal/Center News* and *Metal Statistics* to Fairchild Publications, a division of Capitol Cities Broadcasting Corporation. It retained its other printing and publishing operations in New York City, the big newspaper printing plant which it had built in New Jersey, and *The Oil Daily*, which it continued to run for a few years until it was sold to the *Financial Times*.

It was the end of an era. Fairchild merged its own weekly, the Chicago-based *Metalworking News*, which it had run for about 20 years, with *American Metal Market*, and continued publication of the latter without interruption. In some ways, its character has changed, but its basic purpose does not seek to be greatly different from that which prompted its creation at the end of the last century. Archer Trench used to claim that the paper was "as vital to the metal-buyer as the ticker is to Wall Street", and that still sums up the position for, just as the stockbroker depends on the ticker, so the buyers of the nation's raw metallic material depend for much of the information necessary to run their business on *American Metal Market*: that the vast majority of them subscribe—and at a pretty high rate—is proof of the paper's stability and accuracy.

Certainly, Pat Walker, who has been editor since 1974, is no advocate of major change unless it can be shown positively to be in the best interests of the paper. She, like the newspaper she directs, likes to stick to facts. She has the ideal background and the right personality for the job. After graduating at the University of Illinois, she took a Master's degree in journalism at Michigan State University, and then applied theory to practice on the *Chicago Tribune*, working on general news as well as for the society page. She joined *Metal-*

working News in Chicago some 10 years ago and moved to New York to work under Robert Mastro, an editor with a reputation for being a strict driver. Shortly after the merger with the *American Metal Market* company, she became managing editor and had much to do with the complexities of the transition before being promoted to the top editorial job in 1974, winning out over several male competitors.

Inevitably there were a few murmurings of "it takes more than a pretty face . . ." (and it is a very pretty face) but they soon dried up when she demonstrated quietly, but firmly, her professional ability and her capacity for responsibility. Today, she directs a staff of more than 50, devoting much time to planning and co-ordinating assignments to ensure a flow of good copy. She concerns herself with the Forums, acts as chairwoman and interviewer at such meetings, and often speaks at other national and international metal group gatherings. She does a fair amount of travelling—to see district reporters, for metal plant visits and to conventions—but is most at home behind her editorial desk.

"Her main objective", according to one of her associates, "is to report promptly and accurately news and prices without stress on opinion or lengthy interpretive pieces unless they are vital to the story. She is excellent in guiding the training of new reporters: her general policy is to have people who are prepared to cover any metal, rather than to develop specialists too soon. She likes to keep a fair balance between producer and consumer interests. Her main goal is to provide a reader service that presents news about developments and trends which will, in turn, help to guide sellers, buyers, manufacturers, consumers, merchants, traders and others in making their decisions. Pat does like to get intelligence behind the regular news releases—and it is in this way that she develops the interpretive approach when it needs to be done."

A successful daily newspaper, no matter what field it covers, demands a professional at the head of its editorial services. *American Metal Market* certainly has one.

*

From its early days in the First World War, the prime function of *Metal Bulletin* has been to provide the industry with a pricing service.

"Everything that we write editorially is in one way or another an amplification of a message that has already been conveyed by a price," says Trevor Tarring, for many years the paper's non-ferrous metals editor and today its managing director. "It may be our price—as it would be for tungsten, or cadmium, or scrap, or ferro-alloys, or manganese ore—or it may be someone else's price—such as the LME price for copper, International Nickel's price for nickel, or Alcan's price for aluminium. But basically, all our editorial content has a direct relationship to price and to all those factors that influence price."

The same might be said of *American Metal Market*, or *Metals Week* and its predecessor, *E/MJ Mineral and Metal Markets*. But *Metal Bulletin* claims that there is an important difference, and that difference lies in the considerable degree of judgement which goes into the compilation of its quotations. "For example", says Tarring, "the *E/MJ* United States domestic copper price was based on an undertaking by the American producers to report every ton they sold and the price they got for it, so that a weighted average could be worked out. It was a simple mathematical calculation. On the other hand, there is no commitment on the part of those whom we approach for details of their business to tell us the truth, the whole truth and nothing but the truth. In the ultimate, the figures which appear in *Metal Bulletin* are the result of our own assessment and their publication is our sole responsibility."

Metal Bulletin makes its market enquiries twice a week, on each of the mornings of the day before the paper goes to press. This is no mean undertaking when one considers how many people are contacted and the number of metals covered—and the very many different grades and varieties of some of them. The result is an up-to-date and comprehensive guide to the international market, the paper's quotations being used as the basis of thousands of contracts between buyers and sellers of metals from Europe to the United States, to the Far East and into the Communist countries; anyone who trades with Russia or China, for example, has to follow the *Metal Bulletin* prices which are every bit as valid as, say, the LME price for copper or lead.

The importance of scrap in the metal market was recognized early in the paper's history and prices have been regularly published, not without some opposition from those concerned in this trade who tried for a long time to keep their prices confidential. Today, there are several specialized papers dealing with scrap, but *Metal Bulletin* is still a leading authority to which market men turn for a price guide.

Not all the paper's metal quotations are accepted without question or challenge—and there are some people who will never acknowledge that they use the service that it provides. A favourite story concerns a dealer who went to Canton to do business in antimony and was treated to an outburst about the wicked capitalists of the West and their running dogs, especially *Metal Bulletin* and its prices, which everyone knew were tendentious and unreliable. Eventually, when the tirade came to an end, they got down to business and, after quantities and delivery dates had been discussed, the dealer asked his Chinese client what he had in mind about pricing terms. The reply was immediate: "One per cent over *Metal Bulletin*."

The *Metal Bulletin*'s twice weekly appearance on Tuesdays and Fridays is an essential feature of the metal trading scene. Because it is such an integral part of the market, its influence may not always be apparent, though it is generally recognized that if it did not exist a large amount of international business would need to be done very differently. Its value is reflected in its circulation figures (currently around 11,000, of which more than two-thirds is outside Britain), a relatively high subscription rate, and the fact that, compared with most trade periodicals, an unusually large proportion of its revenue comes from sales to readers and not from advertising. In this respect the paper is in parallel with *American Metal Market*, which has a high subscription revenue as opposed to such periodicals as the advertisement-packed *Iron Age*, whose free distribution in its battle with its twin and rival *Steel* (now *Industry Week*) soared to over 100,000.

Another indication of the value of *Metal Bulletin*'s quotations is the length of the subscription list to the paper's "prompt price" service, under which prices are telexed as soon as they are compiled—and even before they are published—to clients all over the world.

The origins of *Metal Bulletin* go back to the years just before the First World War when an enterprising Irishman, L.H. Quin, was working on *The Ironmonger* as a market reporter. Quin recognized the potential for a London-based international paper specializing in the marketing of all metals and he quit his job in order to start one. Initially, he must have had rather a lean time because his former employers held him to a contract under which he was not to concern himself in any publication which could be regarded as competitive with *The Ironmonger*. But Quin got round this by issuing a series of market "letters" and, in 1914, he also established *Quin's Metal Handbook and Statistics*, which became something of a market-man's bible and was the forerunner of today's *Metal Bulletin Handbook*.

In 1915, almost exactly two years to the day after leaving *The Ironmonger*, Quin produced the first issue of *Metal Bulletin*, its front page liberally sprinkled with advertising. But then came another set-back: under wartime regulations, every increase in a company's profits in the first year of war compared with the last year of peace, were appropriated by the government as "war profit" and, as Quin had made no peacetime profits, he had to give back everything he made. However, he struggled through and was soon reporting prices which had hitherto been regarded as unobtainable.

Soon after the end of the war he had reached a position where he could afford some editorial assistance and, in 1920, he took two young men onto his staff. One was 16-year old Leslie Tarring, a rather shy lad who had ambitions to become a Fleet Street journalist. Leslie's mother had been at school with Mrs Quin and, on her initiative, a place was secured for her son with the object of giving him some basic training and experience in practical journalism ... a stepping stone to a career on a London daily paper. Leslie stayed with *Metal Bulletin* until he died 44 years later.

The other 1920 recruit was Harry Cordero, four years older than Leslie and as extrovert as his colleague was reserved. Of Spanish extraction, he had served in the Intelligence Corps in the War and had picked up a smattering of languages; this was just what Quin wanted as he was anxious to further the paper's expansion in Europe, and Cordero was promptly hired.

The two lads' contrasting temperaments were obviously good for the paper, which went from strength to strength under the overall direction of Quin, who displayed all the colourful eccentricities of his race. It is remembered that his office had two doors, one of which led directly to the street, and it is said that members of the staff who needed to be disciplined were summoned to his office—never to be seen again! When Quin died in 1934, ownership of the company passed initially to his wife, and management to his son-in-law, F.B. Rice-Oxley. To this day, the Bulletin continues to be a family concern, the present executive chairman, F.L. Rice-Oxley, grandson of the founder, joining the company in 1949.

Quin laid down two cardinal rules for the running of the paper: one was that its frequency of publication—twice weekly from the start—should always be maintained; the other was a high level of editorial content and a fast and accurate service for the industry which it set out to serve.

Leslie Tarring and Harry Cordero, each in his own way, continued to keep these ideals very much in mind after Quin's death. They appeared in the "credits" as "joint editors", though it was Leslie who steered the paper to the presses twice a week. He was a man who stuck rigidly to the ethics of his profession, who did everything that was expected of him—and a great deal more, all within a framework of conventionalism and good taste. He was the custodian of editorial integrity and was largely responsible for building up the paper's reputation for reliability.

Harry Cordero was a swashbuckler, the "front man" of the business who loved travelling and meeting people. He never went into an aircraft if he could avoid it, preferring the more leisurely and comfortable routes by rail and sea. One of his favourite excursions was a quick round of the European metal markets. Starting from Liverpool Street station on the boat train for the Hook, he would visit the metal plants of Holland and Belgium and then move on to Hamburg and Frankfurt, finishing up in style in Paris before returning home, on the boat train again. During the Second World War, when there was a scarcity of news and a limited circulation because copies could not be sent abroad, Harry wrote anything and everything to keep the presses turning whilst Leslie became deeply involved with the Ministry of Economic Warfare and undertook other similarly solid responsibilities. A business of this sort needs opposites—and they were *so* opposite.

When Harry's son Raymond came into the business in 1949 he was given the job of writing about steel—Leslie never really liked steel—and, like his father, Trevor, who joined the firm a few years later, concentrated on the non-ferrous side.

It is said that every newspaper becomes the creature of the man who writes it, and every business becomes the style of the man who controls it. Assuming that there is something in heredity, it is not surprising, therefore, that a man should find it congenial to work in a business which has been moulded by his father. What is exceptional about *Metal Bulletin* is the cohesion through two generations of three families—the Tarrings, the Corderos, and the descendants of Quin, the Rice-Oxleys. More remarkable still is the long partnership between the Tarrings and the Corderos in the paper's editorial affairs—a 58-year record which ended in 1978 with Trevor Tarring's appointment as managing director; Raymond is now editorial director.

The Second World War was a difficult time for *Metal Bulletin*, restricting its circulation to the United Kingdom, and if it had not been for the strong family affiliation it may well have gone under. Post-war recovery was quick, the breakthrough coming when it was able to resume publishing international news for international readership. As mentioned earlier, two-thirds of its present readers are outside the United Kingdom. Today, the twice-weekly Bulletin has two stable companions, *Metal Bulletin Monthly* and *Industrial Minerals*, both introduced within the past 10 years. A book-publishing subsidiary, first started by Quin in 1937, has a long list of titles, including the annual *Iron and Steel Works of the World*, which was an inspiration of Harry Cordero and is now practically a business in its own right.

But price reporting and analysis remain the keynote of *Metal Bulletin*'s operations. The number of sales contracts that have been written on its quotations are incalculable and they are also of infinite variety. One, concluded in 1978, is probably unique. The transaction concerned an arrangement whereby a tenant agreed to lease land from a landlord at a rent based on a certain price quotation published in the Bulletin. Under the terms of the agreement, if the quotation falls below a certain figure, the rent goes down, and *vice versa*. The tenant, it need hardly be said, is a mining company, and the metal to which the rent is linked is one of the more notoriously volatile of all commodities. There should be plenty of fun for all.

CHAPTER EIGHTEEN

A Letter to America

To: Doctor Joseph Zimmerman,
    Vice-President, Miles Metal Company,
    New York. NY 10017.

Dear Joe:

You will remember that earlier this year we exchanged letters about my plans to research material for a book dealing with the achievements of journalism in the field of mining and metals, and the contributions which the media has made to the development of the industry. You were kind enough to wish the project well and your warm words have been of considerable encouragement to me in a task which has assumed somewhat greater proportions than I envisaged when I embarked upon it.

However, with the generous help and co-operation of many people, both within the industry and among the media which serves it, I was able to proceed at a fair pace until, suddenly things came to a halt. The time had come to write about Dr. Joe Zimmerman and I realized that, whilst our acquaintance goes back more than 20 years, there was a great deal I did not know about your life and work—and, in any case, 20 years is but a brief period in the span of an octogenarian.

We first met, you may recall, when I accompanied Sir Ronald Prain, chairman of the Rhodesian Selection Trust group of companies, to New York in the mid-1950s to help run what we used to call our "informal" stockholders' meetings. When the group moved its domicile from London to Africa in 1953, annual general meetings naturally had to be held in the country of our adoption, but the vast number of shareholders remained in the United States and in Britain. Accordingly, we thought it a good idea to have "informal" meetings in London and New York in order that members who, though unable to exercise voting rights, would at least have a chance of hearing a first-hand report of our affairs from the chairman, and to question him and other executives about our problems and prospects.

You were always an enthusiastic supporter of these meetings in New York, held initially at the Downtown Athletic Club, and later in less "business-orientated" surroundings uptown. Your questions from the floor, and at the Press conference that followed were always pertinent, and your observations and comments when we used to meet together with Sir Ronald in a relaxed atmosphere after everyone else had gone home, were, for me, the highlight of the day. It was then that I had a glimpse of the real Joe—the man with the enquiring mind ... the shrewd metals economist with an instinctive "feel" about a situation ... the newspaperman who knew most of the answers before he put the questions ... the humorist who loved an "earthy" story and had a fund of them for all occasions ... the teacher from whom so many had learned their lessons in the metals business, but who was always ready to be taught ... and, for me, the guide in a market jungle which was more terrifying than any forest in tropical Africa. I like to think that these brief encounters were of great benefit, both to our group and to the metals market as represented by the readers of your paper.

Later, I heard with mixed feelings that you had given up editing and had accepted a senior post in the industry. Fortunately, you did not relinquish the position you had attained as one of the industry's foremost commentators and critics, and your views continued to demand close international attention. With Archer Trench, you also rejuvenated the dying Copper Club, by introducing a Copper Man of the Year award, of which, if I remember rightly, Clyde Weed, of Anaconda, was the first recipient. I am told that attendance has soared ten-fold from the few dozen people who used to meet to celebrate and watch a two-bit magician, and today the Copper Club dinner has become an event almost as important as the LME annual jamboree in London, with metals men from all over the world attending.

But this was all hearsay. For a book of the sort that I hoped to write, I needed to substantiate these things, so I wrote to you to seek your co-operation in ensuring that what might be credited to you and your times would stand the scrutiny of history. You replied with your usual warmth and courtesy that you felt that you were unable to do this for a number of reasons, which you fully explained. It was an unhappy moment for me when I received your letter, though I understood and fully appreciated your reasons for declining my invitation. However, your decision did place me in somewhat of a dilemma: I knew that the mining and metals world would never accept a book of this title which did not include an acknowledgement of the outstanding work of Dr. Joe Zimmerman.

I had therefore to look elsewhere for my information. I approached a number of people, inviting them to contribute a few words for the record about their association with you, together with an assessment of your work and personality. They all responded. Naturally, I cannot publish in this book all that they have said—and I know that you would not want me to do so! I have therefore selected three letters, each from a man who is well known to you: first, a leader of industry who, incidentally, began his career on the staff of a mining paper; secondly, one with whom you worked very closely; and thirdly one who was a contemporary journalist with a paper which, to some extent, was a competitor of the *Daily Metal Reporter*. Their letters are printed below. I think that together they express what everyone in the metals world feels about Joe Zimmerman.

Yours most sincerely,
Arthur Wilson   London, April 1979

*

By Simon D. Strauss, vice-chairman of Asarco Incorporated, New York

My personal acquaintance with Dr. Joseph Zimmerman goes back to the late 1920s when as a youngster I joined the staff of the

227

Engineering and Mining Journal at the invitation of Arthur W. Allen. Mr. Allen, a British mining engineer, had been a good friend of my father in Chile and was perhaps the only non-U.S. citizen to be editor of the paper. I was assigned to work on the reporting of the metal markets and this immediately brought me into contact with Dr. Zimmerman, who was then reporting the metal markets for the *Daily Metal Reporter*.

He was then a relatively young man in his mid-30s but his warm personality and obvious sincerity had already engaged the confidence of the leaders of the non-ferrous metals industry in the United States. These included such giants of the industry in those days as Cornelius Kelly, chairman of the board of Anaconda; Francis H. Brownell, chairman of the American Smelting and Refining Company (now Asarco Incorporated); Harold Hochschild and Otto Sussman, of the American Metal Company (now Amax); and Clinton Crane, president and later chairman of St. Joseph Lead Company (now St. Joe Minerals).

It was obvious to me that these men discussed the state of the metal markets with Dr. Zimmerman with complete frankness; that he gained their respect by treating confidential material as confidential; and that while he retained his own independent point of view in regard to the probable course of markets, he never invented motives for market actions or pressed his own interpretation of the underlying factors behind a specific company action.

As the years passed, these leaders in the non-ferrous metals industry were replaced by others. Dr. Zimmerman promptly established good relations with their successors—including, for example, Louis Cates and later Robert Page of Phelps Dodge; Howard Young of American Zinc; Dr. John Thompson, of International Nickel, Clyde Weed, of Anaconda; Roger Straus, of American Smelting and Refining Company; and countless others.

Indicative of Dr. Zimmerman's strong ties with industry leaders were the retirement parties he arranged for many senior executives, attended by metal producers and consumers alike. These affairs were usually held either at the Lawyers' Club, in downtown New York, or the Yale Club, in midtown New York.

Many companies honour their retiring senior executives with lunches or dinners, but traditionally such affairs have usually been attended by the corporate associates of the retiree. At Dr. Zimmerman's affairs he invited a wide cross-section from industry to take part in tributes to notables in the industry. Inevitably a photographer took pictures of these events—my desk drawers are full of photographs from retirement parties arranged by the indefatigable Dr. Zimmerman.

In the course of these parties Dr. Zimmerman would act as chairman and would call on several of the friends and associates of the retiree to make appropriate remarks regarding the distinguished career of the honoured guest. Also, inevitably, in introducing each speaker, Dr. Zimmerman would tell a joke, usually of a rather racy quality. While his repertory of such stories is large, the frequency with which he organized farewell affairs eventually led to frequent recounting of some of the stories. Such was the affection in which Dr. Zimmerman was held that the guests laughed as loudly at the third and fourth telling as at the first. A number of us would try to find new stories for our genial chairman, to spare the group the repetition of some of the older chestnuts.

When Dr. Zimmerman finally stepped down from the *Daily Metal Reporter* he joined the staff of Miles Metal Company as vice-president and senior industry oracle.

Although no longer engaged in the daily reporting of the market, his communication with senior executives in the industry has continued unabated. He still organizes retirement parties, speaks frequently at industry meetings, and in general functions as one of the industry's most eminent soothsayers.

On the personal side, Dr. Zimmerman's devotion to his wife, Ann, was outstanding. Together, the two of them travelled around the world during his relatively infrequent vacations. Everywhere they went they were received by leaders in the local mining industries with great hospitality. Such is the international character of the metal business that his name is as well known in Japan and Yugoslavia as it is in Illinois or Arizona—as someone who understands the complex and constantly changing ramifications of the metals business.

Joe Zimmerman started out in life as a teacher. Nominally he changed his profession, first to that of an editor, and more recently to that of a metal trader. But, in fact, he has remained a teacher all his life—everyone in the metals industry who knows him has learned from their contacts with him.

*

By Si Wakesberg, commodities vice-president,
National Association of Recycling Industries Incorporated, New York.

I first met Dr. Joseph Zimmerman back in the early 1940s when I went to work for the Atlas Publishing Co. which was the publisher of a string of trade and industrial publications, chief among which were the *Daily Metal Reporter* and the *Waste Trade Journal*.

Joe Zimmerman was the editor-in-chief of all the publications and, as such, exercised much influence in the metals community. He was, at that time, a man of formidable authority on metals, with a reputation as a copper analyst and scholar. He was a professor at the College of the City of New York prior to his entrance into the field of journalism and one could tell, by listening to him, that he was a good teacher. He always dressed very formally in a suit with a vest and wore a Phi Beta Kappa key in his vest pocket.

Dr. Zimmerman was not only the editor-in-chief, but he wrote the market reports and editorials for the *Daily Metal Reporter*. He seemed to have a special "feel" for metal markets, a kind of seventh sense that stood him in good stead in all the years he edited the publications. His reports were read eagerly by members of the metals industry because of his nose for news, and his reputation for honesty and integrity.

He seemed to know every important executive in the metals industry—particularly in the field of copper. Listening to him talk to presidents and chairmen of the famous copper companies of those days was listening to someone who was on a first-name basis with the great personalities of the day.

I still have a copy of the last special issue of the *Daily Metal Reporter*—a slightly yellowing and dog-eared copy of the Annual

Metal and Steel Number, Volume 61 No.8, dated January 13, 1961. On the front page are articles by Roger Blough, chairman of the United States Steel Corporation; Robert G. Page, president of Phelps Dodge Corp.; and Richard S. Reynolds, jnr. president of Reynolds Metals Company. It is indicative of Zimmerman's wide contact with the leaders of the metals industry that he was able to get these people to write for the paper.

Later, when I became the editor of *Waste Trade Journal* and market editor of *Daily Metal Reporter*, I began to appreciate even more his comprehensive grasp of metal markets—and especially his knowledge of the scrap markets. It was unusual for a man so well grounded in primary metals to have such an understanding of what we now call recycling.

Quite often, the United States Government would call upon him for assistance, and he would receive letters or phone calls from agency heads, requesting opinion on subjects relating to the field of metals. On many occasions he went to Washington to advise Administration or Congressional leaders on problems facing the metals industry.

He lectured at Columbia University on a regular basis but unfortunately I never sat in on any of these lectures. He helped arrange meetings at which outstanding personalities in the metals industry participated. Later on, he became active in the meetings of the Copper Club. He received recognition for his work on several occasions and it is with particular pleasure that I recall how he was honoured in the early 1960s by the National Association of Recycling Industries, with which I had become associated.

When he left the *Daily Metal Reporter* to join Miles Metal as a vice-president, he was tendered a farewell luncheon that is still remembered by some of us old-timers. Rarely has there been such an aggregation of metal industry leaders in one room as on that occasion.

In the years I worked with Joe Zimmerman many young men passed through the publishing house in which he became editor-in-chief and were greatly influenced by him, both as an editor and as a man. He was an excellent teacher and those who wanted to learn could absorb much under his tutelage. He was a tough taskmaster, believing that one had to work hard to do a job properly. But, at the same time, he was always ready to "put out", as they now say, and help the young novice by sharing information, showing him how to analyse markets and letting him in on the little secrets of the trade.

Luckily he is still around to talk about metals to those who want to listen. His fantastic repertory of metals history and experience is the best anywhere. I hope, for the sake of the industry, that he finds time to put it all down on paper or tape.

*

By Alvin W. Knoerr, editor of Engineering and Mining Journal *1955–1969, and now with U.S. Bureau of Mines, Washington, DC.*

I met Joe Zimmerman frequently in the 50s and early 60s when he and I made the rounds of all major metal sellers in New York. Joe was seeking information for comment and reporting in the *Daily Metal Reporter*; and I was obtaining sales tonnages and prices on major metals for the price determinations published in *Metal and Mineral Markets*.

Even though our publications were competitors in the sense that they both sought advertising for operating revenue, I never regarded Joe as a competitor, but rather as a respected and hard-working contemporary. Ever since, in our many meetings at major Press conferences held by the large mining companies in the United States and in Canada, I think our relationship has been one of mutual esteem and admiration, as some of our more recent correspondence reveals.

I do know that Joe organized and held monthly market luncheon meetings which were regularly attended by all the important people in the metal marketing field. While I attended only a few of them, I recall such great mining men as Dr. Thompson, of International Nickel, offering commentaries at the meetings—and Joe was famous for coming up with some choice humour on all these occasions. He was regarded with great warmth and esteem by all who attended. Joe is a very fine gentleman.

I think we had a special appreciation for each other because we were digging for information in the metal marketing field, which in those days was a tough, rough and competitive business. Sometimes I got my head taken off by irate metal sales managers for some of the information we published, and I am sure Joe must have had the same experience.

Since Joe was a market specialist and, in contrast, my journalism covered the entire field of world mining, in addition to market reporting, he was by far the superior market journalist because of his wide and intimate acquaintance with mining men on both sides of the ocean. It is fantastic that, at the age of 86, he can still present such a brilliant paper as he did at the BIR* meeting in Geneva in 1977.

---

* Bureau International de la Récupération. Dr. Zimmerman's address on this occasion was entitled "The Copper Industry in Search of a Cure".

CHAPTER NINETEEN

Today's News Today

Whilst the specialized Press—and more particularly the journals reviewed in Chapters 16 and 17—have made substantial contributions to the development of today's highly-sophisticated metals market, those who are directly concerned with buying and selling the products of the world's mines require a far speedier price and news service than any which these papers are equipped to supply. Even an airmailed weekly cannot entirely meet the needs of those members of managements, located in distant lands, who have to make quick decisions on the repercussions of international events. One cannot, for example, expect a big group of copper companies in central Africa to wait five days to hear that Chilean producers have declared *force majeure* on their deliveries, that the United States government has decided to restrict imports of the metal, or that production in another part of the world has come to a halt because of political or labour troubles. The copper market, in particular, is a highly volatile one and producers who cannot keep up with the news soon lose their competitive position.

Most of the top decision-makers in the international metals business depend for much of their daily information on one of the big news agencies, chief of which are the London-based Reuters, a limited company, owned by associations representing newspaper interests in Britain, Australia and New Zealand; Associated Press (AP) of New York, owned by a co-operative of American and Canadian papers, and linked, in its financial service, with Dow Jones, publishers of the *Wall Street Journal*; United Press (UP), a New York privately-owned enterprise with similar markets to AP; and Agence France Presse, a Paris-based, non-profit trust with a state subsidy.

Of these, Reuters is best known today in the mining and metals field, not only through a long association with some of the world's big mining groups, to which it has supplied both basic and "tailor-made" services, but because of its dramatic expansion since the mid-1960s by the application of computer technology. This expansion has not, as might be supposed, come from any increase in its traditional service to newspapers, broadcasting and other public media, but is almost wholly derived from the transmission and sale of information to the business and industrial world. Specialized economic news today accounts for more than 80 per cent of its turnover—and newspapers, which used to sustain the Reuters service, are only the runners-up in its revenue-making.

This situation represents the completion of a full circle in the agency's affairs, for it was on the transmission of commercial news that its founder originally based his business. Born at Kessel, in Germany, in 1816, Paul Julius Reuter began a career in banking, but moved to Paris in 1848, working for a time with Charles Havas, whose news agency was the forerunner of Agence France Presse. After an abortive attempt to establish his own news sheet in Paris, Reuter found his opportunity in the opening of the telegraph line between Berlin and Aachen, followed in a few months by the extension of the French network from Paris to Brussels. A gap of 100 miles between Aachen and Brussels still had to be bridged to link Europe's most important commercial centres, Berlin and Paris. Reuter hired 40 pigeons from an Aachen brewer and baker, and ran his pigeon-post for almost a year until the telegraph wires were linked.

Looking elsewhere to apply his enterprise, Reuter moved to London to be strategically placed for the opening in 1851 of the first submarine cable joining England to the European mainland. London stockbrokers welcomed his stream of quotations from European centres, but newspapers spurned his offers of a general news service and were slow to see the significance of the Dover-Calais cable. Reuter persevered, building up his contacts in the city, but it was seven years before he persuaded the *Morning Advertiser*, and then six other newspapers, to take a fortnight's free trial of his foreign telegrams. *The Times*, secure with its own extensive news-gathering resources, had for years resisted his approaches, but now it, too, agreed to subscribe and the first Reuter credits began to appear on its pages.

Typical of the enterprise of this energetic little man was the way in which he saved almost a day on news from the United States before the Atlantic cable was laid. He arranged for despatches in special phosphorescent canisters to be thrown overboard from the mail-ships when they arrived off the south-western tip of Ireland. They were picked up by cutters and telegraphed to Cork over a line which Reuter had specially laid, enabling the news to reach London many hours before the ships arrived. In this way, he achieved one of his most outstanding scoops, supplying the London papers with the news of the assassination of President Lincoln two days before anyone else in Europe. The mail-boat was already leaving New York when the Reuter man got the news. He hired a tug, caught up with the mail-boat and threw his canister on board.

Within 25 years of its foundation in London, Reuters was a solidly-based institution. At home, the company had by now the backing of nearly the whole of Fleet Street and, under an arrangement with the newly-established Press Association, supplied British provincial news-papers with despatches from abroad in exchange for domestic news. Gradually, Reuter extended his organization throughout Europe, across to America with the Atlantic cable in 1866, and then to India, the Far East, Australia, New Zealand, and South America. Before he died in 1899 he had realized his dream of a world-wide telegraphic news agency.

In the early 1900s, under the leadership of the founder's son, Herbert, the agency's fortunes seemed secure. But the company soon ran into financial difficulties, which were aggravated by the outbreak of the First World War, and when in 1915, Herbert, overcome with grief at the death of his wife, committed suicide, the whole organization was in danger of collapse. It was saved by Roderick Jones, Reuters man in Johannesburg, who returned to London, raised a loan with the help of influential friends, and put the company back on its feet. In the 1920s the agency was probing for new markets and new ways of disseminating news; the answer was found in a revival of the business which made Julius's fortune, the reporting of commercial information. A new department—Reuter Trade Service—was set up and, to achieve speed and simultaneous reception of messages, the company turned to wireless telegraphy. By 1938, short-wave transmitters were carrying the bulk of all Reuters news from London to the rest of the world.

British newspapers first acquired shares in the company in 1925 when the Press Association, representing British provincial papers and Irish papers, took up a holding. They were later joined by the Newspaper Proprietors' Association (now the Newspaper Publishers' Association), representing British national newspapers; the Australian Associated Press; and the New Zealand Press Association. A trust was formed to ensure independence and impartiality, and continues to control policy and safeguard its principles to this day.

Another important development—and of special significance to mining and metals interests—occurred in 1944 when Reuters bought Comtelburo, a private company with a virtual monopoly of reporting prices between London and South America. Grafted onto the British agency's commercial services, an important foothold was secured in South America, leading to a much-improved flow of mining news from that area; it also laid the foundations for the build-up of the present-day Reuters Economic Services.

With Reuters obtaining easier access to the centres of mining activities in the remoter parts of the world—Australia, South America, Canada and Africa—news of exploration, mine development and company results came chattering over the teleprinters in ever-increasing volume and at ever-increasing speed. In addition, there came faster and more reliable news of events which affect the vital balance of supply and demand in the world's metals: changes in government mining policy, "wildcat" strikes, mine accidents, transportation difficulties, fuel shortages and a host of other economic and political factors that influence the production and marketing decisions of the big companies.

The service prospered as market-men in London and other capitals found that they could accept without question the reliability of the news and the professional integrity on which it was based. And abroad, many of the bigger mining companies, which had hitherto been isolated from up-to-the-minute news about industry events in other parts of the world, began to contract for one of the various services on offer, or to draw up a brief for a service to meet their special needs.

In this way, the market was brought to the doorstep of the producer, even though he might be operating many thousands of miles away, in the African bush, or in the Australian outback. Sometimes, indeed, a combination of service efficiency and time-difference tended to make things embarrassing for London-based representatives of overseas mining groups. A story is told of one resident director in London who used to receive telephone calls before dawn from his chairman in central Africa to discuss that day's *Financial Times*—a digest of which the chairman had already had over his ticker service, but before the paper had been delivered in London. There was magic in the way news was flashed around the world, sometimes, it seemed, even before it happened. It was the same magic that had mystified Prime Minister Clement Attlee, who was persuaded to have a tape-machine installed at No.10 Downing Street on the pretence that it would give him the latest cricket scores, and was then amazed and horrified to see it ticking out details of subjects discussed at his Cabinet meetings.

There is no doubt that Reuters and some of the other international agencies played a big part in gradually breaking down the traditional barriers of silence and reserve that had previously existed between industry—and even governments—and the news media. With more business-orientated reporters beginning to staff agency offices in the world's capitals, and obtaining entry to metal markets, close working relationships developed which fostered a much freer exchange of information and better understanding. During the 1950s, Reuters introduced a range of mailed commodity bulletins, including those covering the major non-ferrous metals, iron and steel; each contained the day's printered stories and more lengthy, undated features. The metal news sheets had a wide distribution to the smaller segments of the industry, and more especially to the weekly and monthly metal journals, many of which relied on them for their coverage of industry developments throughout the world. It was a welcome change from the "scissors-and-paste" exercise that had hitherto produced much news for unsuspecting readers abroad—and which, alas, is still practised in some offices.

In 1964, Reuters applied computer technology for the first time to its economic services by processing the New York Stock Exchange ticker signal, transmitting it at high speed over a transatlantic cable and storing the data in a satellite computer in its London office. Subscribers were linked by permanent line to this computer and, by pressing buttons on a desk unit, like a calculator, were able to obtain immediately the latest prices of any New York stock. Three years later, Reuters moved into direct competition with the American agencies in their own territory, breaking old ties with Dow Jones and setting up a network based on New York, Washington and Chicago. The North American offshoot soon developed new financial coverage and services for commodities, including metals, and most of this was monitored in London for world-wide distribution.

Today, Reuters maintains staff at the London Metal Exchange, in Whittington Avenue. Here, the agency men cover ring and kerb-dealing as well as inter-office trading conducted outside the LME's formal sessions, their reports being distributed on teleprinter circuits and to video terminals. The science of communications has developed to a point where news of this sort can be transmitted over coaxial cables at a speed of around six million words a minute. But, for the metal market, and commodity markets generally, it is not so much the fantastic speed that matters as the fact that all parts of the world can receive the same information simultaneously.

One of the latest innovations at Reuters is known as Monitor, which has been described as "the armchair dealer's dream." By the use of video terminals, subscribers have instantaneous access to continually updated prices; at the same time, news summaries and individual stories can be displayed on the screen at the press of a button. The same system also enables subscribers to display their own quotes and market comment to other selected traders.

Monitor was developed primarily for the international foreign exchange market, which had come to move too fast for normal news delivery systems, but there would appear to be no limit to its applications. The ultimate aim of Reuters is to deliver all its information services on one computer video terminal so that the client, with one screen on his desk, will have access to everything that the company produces.

What effect this might have in the long term on traditional markets is difficult to image. It has even been suggested that some market-places could even eventually disappear. "Never in a thousand years", say the die-hards, but there is a hollow sound in such phrases today: the wind of change can spring up quite suddenly, and already perhaps its breath can be felt around the precincts of some of the City's most hallowed institutions.

*

National newspapers throughout the world—even in countries like South Africa where mining is so much part of daily life—do not, as a rule, devote a great deal of space to the industry, except when particularly newsworthy events occur such as mine disasters; labour problems; environmental and other similar issues guaranteed to arouse the emotions of readers; and confrontations with government over nationalization policies, taxation and other fiscal matters.

Apart from the impact that such events may make in the general news columns, the main concentration of those national newspapers in Europe, North America and Australia which today include a regular mining feature, is on two aspects of the industry: the first is the financial performance of the mining companies, and hence their potential for investors, great or small; and the second is the current value of the industry's wide range of products on the world market. There is, of course, considerable inter-action between these two aspects, but in some newspapers the two are dealt with separately, the first by the mining editor or correspondent, and the second by the commodities editor, who often tends to lump copper with cocoa, tin with tea, and tungsten with sugar.

This somewhat narrow approach of national newspapers to mining affairs is understandable, bearing in mind the enormous range of interests that most papers have to cover. For example, an investor living in Manhattan or Philadelphia could hardly be expected to get excited over news of a technological break-through by an American company in Australia, mining iron ore to be sold to the Japanese, unless this can be shown to mean a few dollars more on his quarterly dividend. Those who want the technical details can read them in the next issue of *Engineering and Mining Journal*, or in other mining papers; national and international news must always be angled to readership, and mining is lucky in that it has media that caters for all interests.

An outstanding exception to these generalizations is London's *Financial Times*, a newspaper "much sought after" by industry public relations officers anxious to promote their companies' interests, and which is unique in many ways.

Before the Second World War, London had two financial dailies, the *Financial Times* and the *Financial News*. The struggle of wartime publishing weakened them both. A pooling of resources to meet the challenge of post-war recovery made a good deal of sense and so, in 1945, they merged under the title of the *Financial Times*. The firm guidance of a brilliant editor, Sir Gordon Newton, led the paper into a golden age as it cast off its rather stuffy "city" image and emerged as a vigorous and wide-ranging international newspaper, catering as expertly for the theatre, literature, art and sport as it does for industry and commerce.

In its mining coverage, the new-look *Financial Times* was lucky to inherit the services of a man who, already before the war, had steeped himself in mining lore. Leslie Parker, a north London boy, had joined the *Financial News* in 1929 at the invitation of its editor, Oscar Hobson (later Sir Oscar, and in post-war years city editor of the *News Chronicle*). It was hardly a propitious time with the world staggering under the great economic depression, but Leslie won his early experience in financial journalism the hard way and battled on until 1934, when his life took a twist which was to provide the groundwork for his later career as a specialist in mining journalism. He left the *News* to become a contributor to the weekly *Investors Review* and, at the same time, took on the task of editing *Sturzenegger's, the Rand Gold Mines*, an annual investment review with some unusual features. *Sturzenegger's* was famous in its day for its encyclopaedic information about the financial and technical progress of every important South African gold mine and was a boon to a generation of stockbrokers who had little else to guide them in their investment decisions in a complex and ever-changing industry.

In 1939 Parker joined the *Financial Times*, but the war soon interrupted his career and when he returned from the Army he found himself one of a staff of two assigned to cover the mining scene for the paper, now merged with the *Financial News*. Adopting the pen-name "Lodestar", he wrote a weekly column which first appeared under the rather unfortunate title of "Mining Tailings" and later changed to the more suitable "Mining Notebook". From the start, "Mining Notebook" was essentially forward-looking. Instead of peppering the reader with share tips like so many of its contemporaries, who were never half as well informed, it set out to guide investment thinking in the right direction. Furthermore, with Parker's pen, it became an influential champion of the mining industry as it came increasingly under attack from environmentalists and all those opposed to the profit motive in mining, or in anything else.

Parker held the post of mining editor of the *Financial Times* for 20 years, which in itself is something of a record in daily newspaper journalism. When he retired in 1975, a profile in *Optima*, the Anglo American group magazine, captured his—and his paper's—essential qualities. The article laid stress on the fact that his main attacks on behalf of the industry were reserved for those politicians who were not only ignorant of the special problems of mining, but who were also not prepared to learn. When Canada swung from a policy of encouraging a healthy mining industry to one in which federal and provincial governments vied with each other in tax-gathering, Parker's "Mining Notebook" headed his comments, "You find, we grab". He called British Columbia's proposal to levy royalties on mines—with little regard for the operating costs of lower-grade producers—"turning ore into waste rock", and said it was "a form of fiscal lunacy which even the so-called under-developed countries have mostly seen to be erroneous". At the same time, he saw no virtue in the sort of inverted racialism which seeks to protect Black developing countries from constructive, if adverse, criticism—simply because they are Black.

He travelled the world extensively in search of the truth about mining. Australia was always a favourite country, and Perth a much-loved city. The great revival in mining exploration "down-under", with its many successes and many more disappointments, always sent Parker hurrying to the scene. It was a journalist's paradise for him as there was so much to write about and so many people to meet. He was keenly disappointed when the Australian government of the day turned to policies which he believed were meddlesome and which slowed down the country's development of its mining potential.

Parker's partnership with his colleague and successor, Ken Marston, with whom he worked for longer than either of them care to remember, was always a happy one, marked by the professionalism of both. News never accrues evenly in daily journalism; papers tend either to be swamped with copy or, to use a "Parkerism", they are "bereft". Yet, even in periods of intense pressure, when company announcements arrived nail-bitingly close to deadline, the mining editor's staff were never allowed merely to "sub-down"; every item had to be carefully assessed and crisply written, with the right emphasis on the important points and, as far as possible, with an element of editorial comment. Above all, the report had to be accurate: a slip in the spelling of a chairman's name destroys in the reader's mind the validity of what might otherwise be a carefully-reasoned comment on a company's performance and prospects.

When Parker retired as mining editor there were the usual mumblings into glasses at farewell parties about being "sadly missed". But no one can be missed if he leaves a legacy of this sort, and no one of his breed can lay down his pen for good. For a year or two he continued to be "Lodestar" and, to complete a circumvolution which encompassed so much, he applied a mixture of his shrewd observation, wisdom and wit to the mining page of the *Investors Review* to which he first contributed some 45 years ago.

In any list of writers who have made a significant contribution to the mining industry, the name of Leslie Parker stands high.

CHAPTER TWENTY

Mining's Public Relations

Whilst the history of mining journalism has been marked by the contributions of some outstanding men who began their careers in the industry and thus had a special appreciation of its problems, it is only during the past 25 or 30 years that relationship between the world's mining companies and the national and international media has been really close and that the mutual benefits of such relationship have become apparent. At the technical level, there has always been a fairly good flow of articles to the Press from within the industry—encouraged, no doubt, by a recognition, both by companies and individuals, of the need to develop a system of industry communications other than through the channels of the professional institutions—but for generations, the idea of sharing information with the financial world, or with the public at large, was quite outside the thinking of those who controlled the mining companies.

It must, of course, be remembered that throughout the 19th century the overriding philosophy of most industrialists, including mine owners, was to extract the greatest amount of profit from their enterprises in the shortest time. The wages of employees were the minimum which would ensure the continuity of a sufficiently large workforce and, with some exceptions, other employment conditions were at a similarly low level. The industry's obligation to government was confined largely to paying its taxes and conforming with safety regulations and other legislation, little of which could really be regarded as restrictive; mining companies accepted few responsibilities for the local community in which they operated; and they did not demonstrate much concern for the industry outside their own environment and their own special problems.

All this has changed in recent times. Today, mining is aware that it has responsibilities extending far beyond the original concept of providing good returns for its owners. This is also true of most other large industries, but it is particularly so of mining, which so often finds itself the predominant undertaking in a territory and the trigger for much other commercial enterprise. Industry's acceptance of its wider responsibilities has come about as a result of two main factors. First, it is, to some extent, a natural development of the growth of industrial power. Secondly, it is the product of the force of public opinion—the determination of people of all races who are associated with the industry to ensure better conditions for themselves and their families. To this must be coupled the desire of government to take a larger share in the ownership and direction of industry and the orderly development of national resources, and the resolution of local communities to obtain the maximum benefit from having a mining operation set down in their midst.

More than 20 years ago, Sir Ronald Prain gave a series of lectures which have had a considerable influence on the development of management thinking on company responsibility. Speaking to students at the Royal School of Mines, he identified seven main responsibilities of the mining industry of the 20th century. The first and most obvious one, he said, remains that which was the prime mover from mediaeval ages until comparatively recent times—the traditional responsibility of management to watch the interests of its owners, a responsibility originally interpreted as being to make sure that shareholders received the maximum return from their investment in the minimum time. The fact that shareholders' interests might ultimately demand longer-term considerations, such as protection over the life of a mine and the most economic recovery methods, may never have occurred to these managers, but such considerations certainly have to be taken into account today.

The next obvious responsibility is that towards employees. Management today recognizes that it is not sufficient merely to have employees; it must have employees who are contented and likely to stay. Consequently, companies have to go out of their way to create conditions which will give security and provide amenities as well as existence. This may involve investing shareholders' money in good housing, hospitals, playing fields and a wide variety of other facilities.

Thirdly, there is the responsibility to government. The industry today must have a far closer relationship with government than in the past; indeed, in many countries where the state has taken a controlling interest in its mining operations, management has to accept a situation in which it either works closely with government or it does not work at all. Then, there is the responsibility of industry towards the community, the people of a district or country in which mining operations are located. Some of the people who live there may do so in order to supply the industry with goods and services, and the consequences of company policies upon this section of the community must always be carefully considered. A modern mine, with all its technical complexity, set down in the middle of a developing country, can create many problems. For example, it can cause considerable disparity of wages between those who work for the industry and those who do not; it can disturb the normal balance between the numbers who live in towns and those who live in the rural districts. It can be argued that this is a responsibility of government but, with the interests of state and industry so closely interwoven, it is inevitable that mining management is concerned in such problems and its expertise is often pressed into service to find ways of redressing the economic imbalance which its operations so often create.

Another responsibility of mining which has been highlighted in recent years by the search for market stability, is that to its customers. Again, speaking of olden times, it was probably natural that management should think no further than getting the highest possible price for its products at any given time. It is now accepted that the highest price is not always the best price, because industry today, as in other responsibilities, has to take a longer view than in the past. The modern world has not discovered, among all its other discoveries, a way of eliminating booms and slumps in commodities and, as a result, conditions still occur from time to time when a commodity, being in short supply, can be sold for prices which bear no relationship to its intrinsic value or its cost of production. Enlightened management does not accept automatically that the price created so ruthlessly and efficiently by the laws of supply and demand is necessarily the price which the industry must ask of those customers whom it would expect to retain as customers over the years: artificially high prices can do the industry much more harm than good in the long run.

A considerable change in attitude has also been detected in recent years by the acceptance of the responsibility that any one section of the mining industry has to the whole industry throughout the world. Mining—certainly mining for metals—produces commodities which are in international demand, and consequently the actions of any one part of the industry cannot be viewed in isolation from those of others engaged in a similar type of extractive operation elsewhere. Today there are international organizations, trade associations, and even inter-governmental organizations, which enable consultation to take place between these various sections of the industry—and they are always very busy. Modern practice is also for the fullest possible disclosure of economic facts, such as production costs and expenditures. Enormous progress has been made since Rothwell attacked the reticence of the mining companies in the first edition of *The Mineral Industry* in 1892, when he declared " . . . secrecy in administration apparently induces, or is induced by, extravagance in management . . . whatever may be the reason given for a refusal to furnish stockholders with such detailed information as is necessary to enable them to understand how the trustees, the directors, are administering their trust, the fact is absolutely certain that secrecy always covers extravagance, and not infrequently dishonesty."

The seventh and final major responsibility of the mining industry, according to the Prain credo, is that which recognizes the necessity to provide for the future. Obviously, this is an obligation which any industry must do its best to discharge, but nowhere is it more important than in mining which is concerned with exploiting a wasting asset. Inefficient usage today, or thoughtless policies designed to take advantage of the short-term position, may seriously prejudice, or even endanger, the future, not only the future of the shareholders and the

247

employees, but the future of countries and industries. In fulfilling its responsibility to the future, the industry needs not only forward-looking management, but high standards of research and exploration, and vigorous programmes for both. To the research man, all waste is opportunity, and this definition becomes more competent as time goes by. If the industry spares its effort in this field it will be failing in virtually all the other major responsibilities that have been detailed. Similarly, with geological exploration: sooner or later mineral resources become exhausted, and it must be part of modern managerial philosophy to divert a proportion of current profits to the purpose of discovering new orebodies wherewith to perpetuate the production of metals. In an expanding industry, the target is not just the reproduction of the industry at current rates of operation, but an expansion which will meet increasing demands.

*

This somewhat lengthy recital of mining's major responsibilities may at first appear to have little relevance in a book whose purpose is to record the contributions which the mining Press and the media generally have made to the development of the industry over the years. In fact, there is a very close relationship, the link being provided by the adoption by the industry of a new service of communication, which is today internationally known as public relations.

When, gradually over a period, mining companies began to accept some of the responsibilities that have been detailed, they also realized the importance of making their new policies and attitudes known, not only within the community in which they carried out their day-to-day operations, but among a wider public whose opinions and reactions could have a direct influence on their interests. This realization did not, of course, come to all companies at the same time—indeed, there are still mining companies in the world today which are reluctant to accept that their affairs are legitimate subjects for public disclosure and debate—but, generally speaking, consciously or unconsciously, the industry became less introverted and much more inclined to communicate news of its achievements, plans and problems to a considerably wider audience than in the past.

There are many reasons for this change of outlook, which became particularly apparent soon after the end of the Second World War. High on the list was mining's need for bigger and bigger investment for new enterprises in a world in which competition for capital was becoming more and more intense. The cost of bringing new mines into production was escalating rapidly as ore-grades diminished and shafts had to be sunk deeper and deeper into the earth. Open-cast methods for massive low-grade mining for copper, for example, opened up new opportunities, but capital costs of the immense earth-moving equipment and treatment plants soon came to be beyond the capacity of any single mining company, or group of companies, and big financial consortia had to be assembled to raise the money. Labour costs were also rising at an alarming rate, particularly in Third World countries as local employees, following the example of their expatriate colleagues, formed their own trade unions and pressed for higher pay, opportunities for advancement and better living and social conditions.

Moreover, whilst British Prime Minister Harold Macmillan had not yet shaken the colonial world with his "Wind of Change" speech to the South African Parliament, some industry leaders elsewhere on the African continent and in other parts of the globe were deeply conscious of the fact that sooner or later the aspirations of native peoples would have to be satisfied and that ultimately they would not be appeased by anything short of both political and economic independence. Several decades earlier, Herbert Hoover had been one of the first to predict that the role of mining companies operating in foreign lands would undergo considerable change as those countries developed. Whatever the exact nature of the change would turn out to be, it was obvious that the most important of the various factors that would combine to determine the new pattern would be the relationship between the companies and the governments of those countries in which they operated, and in particular the increasing trend towards state ownership and control of natural resources and major industries. Those mining leaders who had the vision to foresee such change strove to establish an identity of purpose with the peoples of the emerging nations as a matter of enlightened self-interest, and began to orientate both their expatriate employees and their shareholders to the new circumstances.

To help tackle these strange new tasks, which had never been part of the curriculum at the mining schools, some of the world's larger companies took onto their staffs a number of men and women who had a professional knowledge of the media and an understanding of those sectors of the public with which it was important that the industry should enjoy good relations. To many traditional mine managers who regarded their job as beginning with the production shift and ending with the despatch of a train-load of metal from the mine, this new approach to community problems at first made little sense. But in the higher echelons of management of the bigger groups, technical expertise was already being augmented by a new breed of administrators who accepted that they had to be just as concerned with the local political climate and the behaviour of the world commodity and share markets as they were with what was happening on their own properties. These new members of management were among the first to recognize that professional assistance was needed to explain company policies, interpret the actions of management and foster a climate of mutual understanding and confidence between the industry and its various publics in which operations had the best chance to continue to develop and prosper.

Inevitably, a certain amount of suspicion and scepticism surrounded the early work of the mining companies' public relations men. Like most things new—in fact, only the name was new, the idea behind its practice being as old as time—it was attacked, both through ignorance and fear. For some, it seemed to carry sinister implications of "hidden persuasion" and propaganda; one of the most loathed figures in the Nazi hierarchy in the late war had been Dr. Josef Goebbels, and his evil genius for mass persuasion had not been forgotten.

One of the first tasks of the mining PRO was to establish a closer relationship between the industry and the national Press. Few newspapers—of which, as has been seen, the London-based *Financial Times* was a notable exception—included on their editorial staff anyone with a special knowledge of mining. News from the companies—and it was always very sparse—was treated in the world's Press with little understanding, and very often the main point of a statement or financial report was either lost or misinterpreted. The companies were, of course, as much to blame for this state of affairs as the newspapers, many announcements emanating from the secretary's office being couched in terms which the layman could never hope to understand without interpretation. Another problem was the immense distances which existed in many cases between the mining properties and their headquarters offices, and the newspapers.

Among the first of the mining groups to tackle its Press relations problems in a professional way was the Anglo American Corporation, the giant South African gold producer with direct and indirect interests in a score of other metals and materials, including diamonds, uranium, copper and nickel, coal, platinum, tin, potash, asbestos, iron ore, lead, zinc, manganese and wolfram. In the early 1950s, its leaders in Johannesburg—among them Sir Ernest and Harry Oppenheimer, and their "right hand", R.B. Hagart—were concerned about the generally poor, and all-too-frequently, misinformed editorial coverage that their activities were being afforded, both in South Africa and in Britain, one of the major sources of capital for development. A small public relations staff, each member with practical newspaper experience, was recruited in South Africa and, in London, the group's directors appointed the editor of a Fleet Street newspaper to take charge of its public affairs in Britain.

It was particularly important that Anglo American policies became better understood in the financial world as the group was heavily engaged in raising huge sums of money for the development of the Orange Free State gold mines. At this time the industry in South Africa was not well served by its national Press and few papers could boast a mining specialist on the staff.

In London—save again for the *Financial Times*—mining topics were handled by any member of the city editor's staff who might be available at the time. Few financial writers knew anything about the economics of mining; it is doubtful if any of them had ever even seen a mine—except from a "safe distance". Yet the City was buzzing with unsubstantiated stories about the discoveries in the Orange Free State

251

and the possibilities of development unprecedented in the history of mining. Reliable news was sparse, informed comment non-existent, and rumour was fanned by stockbrokers who had little more information than the newspapers, but were anxious to take advange of what appeared to be the beginning of a boom in gold share prices. The result was a chaotic share market which was doing Anglo American, and the mining industry generally, no good at all.

What the Corporation's public relations men did in these circumstances was a model of sense and sensibility. Realizing that the past reticence of their companies to take the Press into their confidence was a contributory cause to the current unsatisfactory situation, they persuaded their directors to adopt a policy of "full disclosure". Background information about the group's activities and interests were spread far and wide; newspaper correspondents were invited to informal discussions with senior officials from the chairman downwards; editors were briefed about top-level policies; financial news was presented in a manner in which it could be easily understood and analysed; annual and quarterly reports were overhauled and their main features summarized.

Gradually things came right. None of the confidence proved to be misplaced; "off the record" information was respected; the newspapers were as keen to learn as the mining companies were anxious to teach. Perhaps the greatest inspiration was to invite a group of a dozen London financial editors to fly out to South Africa and Rhodesia to visit the properties and see for themselves the magnitude and complexity of modern mining and to obtain an insight into the problems. Again, there were no holds barred: it was "full frontal disclosure"— even the touchy problems of the employment of African labour and the big wage differentials between Black and White workers were laid bare. Today, more than 20 years after the first Press visit, Anglo American still invites parties of newspaper correspondents from all over the world to visit their mines and to discuss their plans and problems. Their older public relations men can look back with amusement to the days when the city editor of one of London's mass-circulation dailies described Sir Ernest Oppenheimer as "not the sort of chap one would want to meet on a dark night in the financial world"; he had, of course, never met him in any circumstances, but after an introduction had been arranged and views had been ventilated, he became one of Sir Ernest's greatest admirers.

The rest of the industry has benefited immeasurably from Anglo American's enterprise and trust, which has contributed substantially to international awareness of mining's place in modern society, and, through the world's newspapers, many of which now have their own mining specialists, its complicated economics and investment problems are better understood.

The 1950s also produced an interesting example of co-operation between mining companies and a publisher which initially helped to ease an unsatisfactory position in the gold share market and has continued to provide a useful service to stockbrokers and others concerned in the progress of the South African mines.

Judged by many standards, the post-war record of South African gold mining had been good: production had risen steadily, new mines had been developed, or were in the process of development, and many of the old ones continued to exhibit surprising resilience. The technical efficiency of the industry was high, and in many cases profits from gold were being helped substantially by new revenue from uranium. Yet, despite all this, there had been a progressive fall in the level of gold shares and they had become more prone than most other investments to be evaluated on hunch or hearsay rather than on facts and figures. In part, this was due to the traditionally speculative nature of what was then known as the Kaffir market; in part, to the highly technical nature of gold share valuation; and to a confusion in the public mind outside South Africa between what appeared to many to be a misdirection of the country's racial and foreign affairs, and the essential soundness of its economy. It must be remembered that in these days stockbrokers did not have the benefit of advice from mining engineers and economists, nor, indeed, from anyone closely associated with the industry, and several years were to pass before some of the larger firms began to employ their own specialists.

The only reliable guide to the market was *Sturzenegger's, the Rand Gold Mines*, already referred to in the preceding chapter, and which was still being edited by Leslie Parker of the *Financial Times*, its owner having retired to his native Switzerland. A limitation of *Sturzenegger's* was the fact that it was published only once a year, and although there were facilities for ardent students of the market to keep their copy up to date by quarterly supplements, few of them bothered to do so. The London *Mining Journal* became interested in the possibility of developing an independent review, based on the quarterly reports of the companies; in 1957 an arrangement was made with Sturzenegger and *Mining Journal's Quarterly Review of South African Gold Shares* was launched.

The first purpose of the *Review* was to collect and tabulate all the published information concerning most of the 60 or more gold mines in South Africa in such a way that the investor could see at a glance all the relevant facts and figures concerning each mine and its performance over the previous two or three years. The second purpose was to provide some interpretation of the data and to offer comment. To make this comment as objective as possible, one of a panel of four London mining journalists submitted his comments on each mine, as did a well-known Johannesburg mining share analyst. These two viewpoints were synthesized and, where necessary, reconciled by the editor of the *Review*, who was also *Mining Journal*'s financial editor. The comment was updated in each issue.

Since its inception, the *Review* has appeared every three months, between two and three weeks after the publication of the companies' quarterly reports, and is sent to its subscribers by airmail. The pattern of its compilation remains substantially the same as in 1957 and its comment is still provided by a panel of "experts". It is, of course, impossible to assess the degree of influence that it has had in stabilizing the gold share market but it certainly has fulfilled its purpose of supplying facts for those who deal upon that market, instead of the mass of fiction and rumour which surrounded its activities some 20 years ago. Since 1974, when a strong interest in South African gold shares developed in the United States, the *Mining Journal* company has extended its service with the publication of a monthly airmail newsletter for the North American investor.

*

Whilst the first concern of those who directed the mining companies' public relations effort in the 1950s was to establish a close working relationship with the national and international Press, it was not long before their activities began to extend to the other fields of responsibility which the industry generally had begun to accept. Throughout the mining world it began to be recognized that there was no longer a choice as to whether a company or group of companies should "have public relations"; the only real choice was whether its relations would be good or bad; either way, they would substantially affect the long-term future of the company. Inevitably, because its first application, as a planned operation, had been in the newspaper field, there were some who confused the functions of "public relations" with the better-understood practice of "publicity". Perhaps the most important difference between these two operations is that "publicity"—defined simply by the Oxford English Dictionary as "the business of advertising, both goods and people"—is very much a short-term effort, primarily designed to give quick results; it frequently seeks to impress through sensationalism and has little lasting value; it can also be extremely costly, and, if applied without careful supervision and appreciation of its inherent "boomerang" effects, it can also be very dangerous. Public relations, on the other hand, is a continuing operation, working within the framework of a well-defined policy, and whilst it may frequently call upon some of the tools of publicity to achieve certain objectives, its strength is mainly derived from the human resources within an organization rather than expensive aids from outside.

No two public relations practitioners will describe the functions of their profession in the same way. The (British) Institute of Public Relations has spent much of its 30 years' existence in thinking up definitions. Bernard Levin, the Fleet Street columnist and TV entertainer, has spent much of his time demolishing them. As long ago as the mid-1960s he was writing:

> The sole objection to Mr. Smith by Mr. Neil is that his is the "wrong image" for the Conservative Party. Now is it not time that this rubbish was knocked firmly on the head?
> The "image" in this sense, was coined by the public relations trade,

which more than makes up in fraudulence what little it lacks in incompetence, to act as a substitute for value. Thus there was recently a suggestion that the theatre industry in Britain should do something to improve its image; it had not apparently occurred to those responsible for the plan that what they should be doing is improving the theatre, not the image.

Similarly, the French Government has just appointed a charming young lady to improve the image of French tourism; there is no intention, apparently, of preventing French hoteliers and restaurateurs from swindling their customers, but there is to be a campaign, regardless of expense, to pretend that they don't.

Mr. Smith, by this sort of reckoning, may be an excellent and vigorous candidate and an efficient and hard-working member. But he has "entirely the wrong image". That is to say, he appears to be what he is, and he is what he appears to be.

Levin is right, and wrong—as he is about most things. Public relations is, or should be, concerned with representing an organization to the public for what it is; to tell anything but the truth about it is not only immoral, but could be disastrous. It is no function of PR to "create images"; any attempt to do so is self-defeating as it immediately arouses suspicion that the present image is tarnished and needs gilding. The pioneer public relations men in the mining groups did not embark on "image-making": they merely made sure that newspapers had the basic facts on which they could form their own judgements and express their own comments; any "image" which might have emerged was the result of the papers' own assessment and not contrived by a scheming PRO. It would be foolish to pretend that PR was, or is, free from malpractice; but also, from time to time, doctors are struck off the medical register, priests are unfrocked, stockbrokers are hammered, and company directors are called into account. Public relations needs a code of conduct as much as any other profession, and if its various institutes can set its standards high they will justify their existence.

The mining industry has, by and large, been fortunate to have been served by public relations men and women who have conformed to a good standard of professional ethics and who have been flexible enough to apply their expertise in a variety of fields. They have, for example, played an important part in employee relations by providing the necessary back-up to companies' personnel policies and, in many cases, by establishing two-way communications between management and labour outside the formal negotiating machinery. Through magazines, newspapers, film, and other media, they have helped to bridge the gap between those who give the orders and those who carry them out, and in some parts of the world their publications and "visual aids" have performed a useful function in keeping employees aware of changing political and economic circumstances affecting the future of their jobs. In shareholder relations, they have helped the companies discharge their traditional responsibilities to their owners, not only in ensuring that the maximum information is given to those who invest in the enterprise, but by seeing to it that the information is presented in an attractive and readable form.

A less-obvious, but equally important, function of public relations in the mining industry stems from the fact that for generations the larger companies carried on their operations in colonial territories and in foreign lands outside the control of their own governments, and subject to the laws of the host country. This situation has changed dramatically in recent years with the growth of state ownership and control in so many parts of the world, and more particularly in the developing countries. But, for years, one of the most important activities of public relations was to assist management in creating and maintaining a relationship with government, leading to a closer understanding and appreciation of each other's problems. For the mining companies, this meant, among other things, co-ordinating their social and economic policies with those of government; for government, it meant learning the complexities of a sophisticated modern industry and adapting its policies to provide for the greatest benefit to accrue to the nation without interfering with the natural growth and efficient management of the industry. In particular, it was important that there should be agreement on such matters as the basis for levelling taxation and collecting royalty payments, labour relations, and on marketing policies in so far as they might affect prices and production control. Naturally, the PRO did not go about such tasks with a label in his lapel, proclaiming his identity and purpose. The "lobby" function, if one might call it that, was a matter of high diplomacy and exercised, on the company side, by the most senior executives in the industry; the PRO—except when he was a member of management in his own right—worked behind the scenes, accepting that in the ultimate no industry, no company, no organization of any sort, can have anyone other than its chairman or president as its top representative in this field.

Today, in the Third World, and in certain other countries where the state has become the owner of its mineral rights and mining operations, the role of public relations for the industry has become inseparable from that carried out for government, or for the national effort generally. It is quite impossible, for example, to think of mining in Zambia, in Chile, or in Zaire without immediately associating the industry in those countries with the policies of the governments which control it and with the political climate in which it operates. Internally in such countries, PROs may well continue to do a useful job in interpreting the industry to the general public, but externally they must needs speak with the authority of government as well as on behalf of the industry.

Public relations in the mining industry has undergone a number of other changes in recent years as the industry itself has changed and life has become more complex for those who are concerned in its management. The web of red tape spun by government departments, the regulations of national stock exchanges and security exchange commissions, anti-trust laws, monopolies commissions, environmental restrictions, and the mass of rules and "guidelines" issued by such bodies as the EEC, make the job of running a mining corporation—and consequently the job of its PRO—more and more difficult. Today, the media itself is more complex, with television, radio and film competing with the Press and offering new opportunities for the PRO.

As far as its association with the Press is concerned, mining industry public relations appears to have turned full circle, reverting largely to PR's original role in the commercial and industrial world of being the dispensers of official information, such as annual reports, chairman's statements and announcements about development projects and investment activities. In these days of specialization, most national and international papers have their own mining correspondents and financial writers who are well versed in the affairs of the industry; there is consequently less need for the "educative" function which was so important for the mining houses to pursue a quarter of a century ago, though there is a continuing need for the interpretative function.

The job of the PRO today is therefore much more routine and formalized. Because the interests of the owners—be they governments or private shareholders—must still come first, and the requirements of stock exchanges and other institutions have to be observed, care must be taken with the timing of important announcements; moreover, it is usually in the best interests of the companies that such announcements should, whenever possible, be issued simultaneously in the world's financial capitals. At times, such practice mitigates against the traditional spirit of competition which prevails in the newspaper world. Most editors accept the companies' position in this respect, but it is, for example, exasperating for a correspondent to obtain interesting information on his own initiative only to be told, when seeking confirmation, that this is to be the subject of a general announcement in a day or two when all the formalities have been concluded.

John Carrington, for 44 years with the *Northern Miner* in Toronto and 29 years its editor, highlights the problem. "The *Northern Miner*", he says, "was nurtured in the tradition of getting there first and we still try to cultivate that spirit among our staff. But it is a different ball game now. The securities commissions have put a damper on news gathering, and I doubt if we shall ever again get a clear beat on such major stories as we did, for example, with the Brunswick discovery or the Texas Gulf find. Nowadays, companies are enjoined to make a public statement as soon as they have any news of major significance. For a weekly paper such as ours, this means that there are five other days in the week when the dailies can beat us to the draw. I don't think the public's best interest has really been served by this regulation, and we see numerous instances of companies withholding news because they say they have to hold it until they can let all their shareholders know, or some such excuse. However, I am under no illusion that the clock can be turned back, and we feel that we have a job to do by more interpretative reporting and covering the news in greater depth than other publications."

The mining industry has itself become a publisher of some size and importance over the years with a steady output of newspapers, weekly, monthly and quarterly magazines for employees and shareholders; and a mass of information books of general and technical interest. Whilst it is difficult to say how far they have achieved their objectives—which have not always been the same in one company compared with another—most of them have been produced to a high professional standard, and some have been outstanding, judged by any criterion. In most of the larger groups, the public relations department is directly or indirectly responsible for all publications, including the physical production and presentation of the annual report. It is in this latter field, perhaps, that the most significant progress has been made, the modern report, with its colour illustrations, diagrams, financial "cakes" and family trees, bearing little resemblance to the wordy, unimaginative reports of yesteryear, which many shareholders found impossible to understand. Today, graphic designers must be beginning to run out of ideas, which is just as well as there must be limits to the growth of such publications, or shareholders may be faced with diminishing dividends to pay for lavish reports which tell them why there are no profits for distribution. The same applies to some other company publications and, in this connection, it is well to remember T.A. Rickard's sharp criticism of "institutionalized" and "subsidized" journalism for, whilst it can be argued that few employee or shareholder magazines compete directly with commercial publications, no publishing house likes to be thought of as mining's "poor relation".

CHAPTER TWENTY-ONE

The Political Football

Although change has been an outstanding characteristic of most aspects of life over the past 25 years, few sectors have been so dramatically affected as the international minerals industry. The extent of this revolution is indicated by one study which concluded:

> The mining industry no longer operates as an economic or industrial unit in a free or open market environment. Rather, it has become a conscious, overt tool for economic, political and social motivations, often used aggressively by developing countries, often subordinated to other facts and needs in the host country.*

Of the various forces and events which have brought about this situation, the most significant is the great surge of nationalism which swept away colonialism in Africa in the 1960s and showed itself strongly in other areas, notably Latin America. In many of the countries that achieved independence in this period, minerals play a very large part in their economy, and one of the first acts of government in these countries has been to assume ownership of all natural resources and take control of the mines. Zaire, Zambia, Chile and Peru were among those which moved towards state-ownership at the earliest opportunity: these countries are major copper producers, and it is interesting to note that government control of copper production has

---

* *The Mineral Industry—an Outlook on Government Policy and Involvement*: Geoffrion, Robert and Gelinas.

261

increased dramatically in the past 15 years or so to a position where state-owned mines account for well over 40 percent of the so-called "free" world capacity.

Simultaneously with these events in the developing countries, there has been a marked tendency in the developed world for governments to concern themselves much more closely with the mining industry. Here, the motivation has been somewhat different. It cannot be argued that mining in, say, Canada, the United States, or even Australia, has been an element of colonialism which must be expunged from "free" society. What, then, are the reasons for this drift towards Third World attitudes?

In the cases of Canada and Australia, one of the principal reasons appears to have been government's fear of foreign domination of natural resources, a fear reflected in measures which have been taken to restrict outside investment in both mineral exploration and the development of mines. In Australia, where there is little enough local finance available for the country's economic growth, a constantly changing government policy has already had an adverse effect on minerals development in the present decade, whilst Canadian mining has also suffered from restrictions imposed on the free flow of capital from the United States, its traditional source of supply, and elsewhere.

A second reason for increasing government control in the developed world, and perhaps more especially in the United States, has been the pressures put upon mining by a rash of social interest groups concerned with the polluting effects of some of the industry's operations and their unsightly appearance in the countryside. Overall, the contributions of such groups have been of much value in bringing about a greater awareness of the need to preserve a decent environment in which people can live and work and enjoy themselves. Unfortunately, the cause has not always been helped by emotional accusations by some of the more ardent conservationists, accompanied by demands which, if met in their entirety, would spell disaster for much industry, have a dampening effect on the national economy and, incidentally, deprive the population of many of those goods and services which contribute to the high standard of living and comfort that it presently enjoys.

To help protect its position in these circumstances of ever-increasing restrictions and controls, the industry has been making use of its various national associations, re-structuring them where necessary in order to pursue an aggressive public relations policy and take on the role of pressurizing government. In the United States, the American Mining Congress (AMC), which has been in existence since 1897, has never been stronger, and never has it been more needed. The association encompasses the producers of most of America's metals, coal, industrial and agricultural minerals, manufacturers of mining and processing machinery, engineering and consulting firms, and even the financial institutions that serve the industry. One of its traditional functions is to bring together mine operators and equipment manufacturers to stimulate technological progress in all facets of the industry, and the magnitude of this operation was demonstrated at the 1978 Las Vegas meeting and show, for which there were more than 25,000 registrations from 49 countries. AMC is fully alive to the fact that whereas in earlier times most of the impetus for technological advancement was largely rooted in economics, today the pressures on the industry stem from political and social causes. This is not unique to mining; there is probably not an industry in the United States that does not feel the increasing intrusion of government into its affairs. It has been said that in recent years Washington has become a lawyer's dream and a businessman's nightmare, and this pattern is being duplicated at state and local levels all across the land.

The very nature of mining makes this phenomenon worse for AMC than it is for other industry associations, and it could never hope to deal with it but for company leaders, who have not only increased their financial backing, but have vigorously stepped up their personal involvement. The magnitude of the task was reflected by Charles Barber, chairman of Asarco, one of the country's major mineral groups, when he told the 1978 Las Vegas convention: "Unless something changes, 10 years or 20 years from now our mineral industry will have disappeared." It is the cost of conforming to the multitude of government regulations, especially during times of depressed market prices, that is particularly worrying the American industry that it will lose its competitiveness in the face of lower-cost overseas producers.

AMC is concerned with every issue confronting the modern mining industry, from occupational safety and health to a national energy policy, and to laws governing the mineral wealth on the ocean floor. Every day, its people work with Members of Congress and their staffs, and with officials in the Administration and independent agencies that impinge on mining. There are some 20 committees, drawn from member-companies, that analyse the potential impacts of new or proposed legislation, devise alternatives and prepare congressional testimony and comments on regulatory matters.

The work is backed by a number of regular publications, including the monthly *Mining Congress Journal* (the one paper that T.A. Rickard excused from his attack on "institutionalized" journalism) with a circulation of 20,000, and the bi-weekly *News Bulletin*, which goes to some 5,000 people in the mining companies, to Members of Congress and the Administration, professors of mining schools, the Press and others. Special brochures are prepared on issues such as ocean mining, and the preservation problems of Alaska. Another brochure, *What Mining Means to Americans*,* is particularly popular among schools and colleges, and a film version of this book, entitled *Within this Earth*, is being shown in commercial theatres throughout the country to an estimated annual audience of about 3½ million.

But AMC's president, J. Allen Overton jnr., firmly believes that personal contact brings the best results. "More than ever," he says, "we have to work on Capitol Hill. There is no substitute for close, continuing, personal relationship between people in our industry and their elected representatives. We will never find friends in Congress if we remain strangers to them. Certainly, the mining industry has a powerful message to convey. The excessive regulation being imposed on it today—threatening to become even more burdensome in the future—means stagnation instead of expansion. It means shortages, higher prices, fewer jobs and more imports. The American people must pay ultimately the price of regulation run riot, and it is they who must rebel against it. If they perceived the truth, they would demand reform. And our task is to advance this perception."

---

* This title is an adaptation of the Mining Association of Canada's *What Mining Means to Canada*, and the Australian Mining Industry Council publishes a similar booklet.

Nowhere have the changing circumstances, under which mining has to operate, been more apparent than in Canada, where the industry, in terms of value and diversity of product, is now the third largest in the world. Here, as in the United States, the change has been marked by the emergence of ecologists, environmentalists, malthusian theories on resources generally, and a growing involvement by government.

As a result, it has become increasingly important for the industry to promote much greater knowledge of the characteristics of mining and the benefits it brings to the country. Quite apart from the substantial programmes undertaken to this end by many of the leading Canadian mining companies, the Mining Association of Canada (MAC) has for some years devoted a considerable proportion of its resources to such a promotion. This has led to projects involving a series of widely-circulated publications, films, audio-visual presentations, educational aids for schools, radio spots and speakers' programmes. Supplementing these efforts, the provincial mining associations have developed more localized programmes, and the Canadian Institute of Mining and Metallurgy, the professional and technical body, whose history goes back to the days of that colourful pioneer of Canadian mining journalism, B.T.A. Bell, is also concerned in a similar type of campaign.

At the same time, MAC has made efforts to develop new consultative mechanisms with government. In earlier days, industry-government relations in Canada were extremely cordial; government departments concerned with mining affairs were small and had a keen appreciation of industry's problems. In more recent years, government's ever-increasing involvement has led to differing philosophical approaches to the whole question of mineral resource development. Relationship with the industry probably reached its lowest point in 1974 when government, particularly at certain provincial levels, introduced a number of policies aimed at collecting more revenue from the so-called "windfall" mining profits. There was also a feeling in the industry that the central government was not paying sufficient attention to mining matters as distinct from energy problems, and there has been particular concern over the increasing cost of compliance with environmental regulations. Whilst most mining companies agree that any development project must be carefully planned, they believe that it should be possible to streamline the innumerable studies which have to be made, and which add substantially to the cost of establishing a new operation, and consequently have a considerable influence on investment decisions.

More recently, relationship has improved, and this better feeling undoubtedly stems from the steps taken by the industry, through its associations and other related groups. A notable example is the initiative taken by MAC in the spring of 1977 to establish a Federal Government-MAC Task Force on Mining. The first object was to re-emphasize at Federal government level the issues and concerns of the mining industry as distinct from energy matters; secondly, the three departments most closely concerned with mining affairs were made very much more aware of mining's problems, particularly in the very sensitive area of taxation.

John Bonus, managing director of MAC, takes every opportunity to hammer home industry's plea for a new Canadian mining tax structure geared to the ability of the industry to compete successfully in international markets. Speaking at the Sixth Northern Resources Conference, in Whitehorse, in October 1978, he pointed out that Canadian mining "faced a bewildering array of taxation across 11 separate jurisdictions"—the result of a lack of federal-provincial agreement on resource taxation policy. It does not seem unreasonable for mining to expect taxation levels in line with other sectors of industry, and Canadian mining men will go on fighting for a better system. Perhaps the substantial lay-offs in Canadian mines and plants in 1978 have helped to make the politicians realize how vulnerable mining can be and, whilst many of the industry's problems are caused by international market conditions, over which it has no control, its viability can be partly protected by more realistic fiscal policies at home.

Top-management people from 34 companies are on MAC's board of directors, representing the great majority of mining interests across Canada. Naturally, these members have their own contacts with government at various levels, but their efforts are in line with the association's general thrust.

Public relations currently absorbs about half of MAC's total budget. "We have every intention of sustaining our efforts in this field," says John Bonus. "However, whilst we should most certainly help ourselves, we should also expect our elected representatives to give us more of a hand than in recent years. Politicians have generally been reluctant to speak out openly in favour of our industry. They have looked upon it as a somewhat unpopular subject because of some adverse publicity we have received in certain areas, and also, I suspect, because although economically strong over a prolonged period, mining has lacked political clout—that is, it has not commanded many votes in relation to other sectors which are more labour-intensive. Yet, politicians must realize that without a strong mining industry, Canada would be a very sorry place in a competitive world and, this being so, they should show more preoccupation with the future of our industry."

In Australia, the mining industry's voice has difficulty in making itself heard, partly because of the vastness of the territory, and partly because, like Canada, it has a multi-tier system of government. The Constitution leaves the states with direct administrative powers over minerals and mining—they issue exploration licences and mining leases, and are responsible for such other matters as environmental control and the internal transportation of minerals. In the two mainland Federal territories, it is the Commonwealth government that exercises these powers. The Commonwealth's involvement in mining arises from its constitutional powers in the taxation, foreign investment and export trade. All this means that the industry must deal with both state and Federal governments and their agencies. Recent administrative arrangements concerning the Commonwealth government established two Federal departments responsible for the industry—the Department of Trade and Resources, and the Department of National Development—thus adding more bewilderment to an already complicated picture.

The Australian Mining Industry Council (AMIC) was formed in 1967 as a forum in which member-companies, which now number about 140, and individuals closely involved in mining, could express their opinions and make them known to the public. The views of the industry as a whole are conveyed by AMIC directly to Federal and state governments and Oppositions, civil servants and to the public, through education and information programmes.

Unlike the American Mining Congress, AMIC does not maintain staff in Parliament House; visits made there by its executive director, G. Paul Phillips, and other representatives are mainly by invitation, or are initiated by them when they have need to consult with a particular minister or group of ministers on specific matters. In recent years they have been invited to meet the Federal Cabinet before the formulation of the Budget.

The mining industry's relations with the public form an important part of the Council's workload—a pretty heavy one in view of the fact that the community generally still knows little about mining and tends to see the industry in terms of issues like foreign investment, uranium and environmental damage. In an attempt to do something about this problem, time has been taken on television to provide the basic facts about mining, and early reports of its effectiveness have been encouraging.

AMIC has an enormous task ahead of it, along a road which will never be easy, but Australia's full potential in the mining world cannot be realized until the problems and aspirations of its industry become better known and appreciated.

CHAPTER TWENTY-TWO

The Mining Press Today

Mining and metals processing is an international business, and as the years go by its Press becomes more and more international, both in its editorial character and in its readership. Of present-day journals produced commercially in this field which can claim a world-wide following, the most important are the monthly *Engineering and Mining Journal* (circulation 23,000), *World Mining* (15,000) and *Mining Magazine* (9,000); the weekly *Mining Journal* (8,000); and *Metal Bulletin*, which sells in the region of 11,000 copies twice a week.

Each of these papers has subscribers in well over 100 countries, both *World Mining* and the *Mining Journal* group having as much as 85 per cent of their readership outside the countries in which they are produced. *E/MJ*'s proportion of non-U.S. readership is about 40 per cent, of which Canadian sales represent a large share, whilst 70 per cent of *Metal Bulletin*'s subscribers are outside Britain—many of them in the United States. Canada's *Northern Miner* can also be found among mining communities in many parts of the world, but few of the other commercial publications that have been discussed in the preceding chapters are much to be seen outside their countries of origin.

Some of the "institutional" journals naturally travel across the globe, seeking out their members in distant lands. Chief among them are *IMM Bulletin*, of the Institution of Mining and Metallurgy, London; *Mining Engineering*, of the American Institute of Mining, Metallurgical and Petroleum Engineers; *Mining Engineer*, of the

Institution of Mining Engineers, London; *CIM Bulletin*, of the Canadian Institute of Mining and Metallurgy; and *The Metallurgist*, of the London-based Institution of Metallurgists.

In terms of absolute numbers, the American *Iron Age* still has by far the largest circulation in the ferrous metals field, with 110,000 copies produced each month for domestic distribution, as well as world-wide and European editions which claim to reach readers in the world's largest plants outside the United States. It must, of course, be remembered that *Iron Age* is a "give away"; there is no price tag and its recipients qualify either as firms, or as individuals with executive and administrative functions in the industry.

A similar system of what has come to be known as "controlled circulation" is operated by *World Mining* which, as mentioned earlier, is one of the "big four" commercial journals serving mining interests internationally today. Anyone outside the United States who can satisfy its circulation department that he is technically qualified in mining, milling, smelting or refining, or is a member of a minerals exploration team, a consulting engineer, or working in a government department concerned in mining, can get a copy post-free. Within the States, one has to have somewhat higher qualifications, distribution being confined to those who have attained the status of executive management, or who are technically employed by one of the big mining or metal processing companies.

The magazine is the successor of the (American) *Mining World*, product of Miller Freeman Publications of San Francisco, one of the best-known publishing houses on the Pacific Coast and which specializes in international journals. Miller Freeman founded his company in 1902 in Seattle, Washington, his first periodical serving commercial fishing interest in the area. But publishing in the Freeman family, which still directs the group, goes back one further generation, to Miller's father, a pioneer journalist of the last century. One of the papers he started, *Frontier Index*, was printed on a hand press and carried across the plains in a covered wagon; it bore the rousing slogan: "Independence in all things, neutrality in nothing".

After his fishermen's paper, Miller Freeman began to give his attention to producing publications to serve the natural resource industries on which the economic growth of the Pacific North-west is based, and as part of this policy *Mining World* was launched in the late 1930s. Ten years later, a sister paper, aimed exclusively at overseas readership, was produced under the title *World Mining*, and for a number of years the two publications ran concurrently. However, international readership—and advertising support—soon outstripped domestic circulation and the company decided to sell the original publication and concentrate on *World Mining*. Today, the magazine includes a fair amount of U.S. news, but its overriding editorial policy is to seek out and report on new developments of international significance to the industry wherever they may be found. The magazine is produced by a multi-national, multi-lingual staff of graduate mining engineers who, from bases in San Francisco, Brussels and Sydney, travel many thousands of miles each year to research their articles. The Iron Curtain has been penetrated several times and the paper has featured authoritative articles on mining in the Ukraine, Uzbekskaya, Kazakhskaya, Poland, Bulgaria, Czechoslovakia and Romania; and in 1978 it undertook a survey of the industry in China.

In passing, it is interesting to note that mining journalism in Russia dates back to the early years of the 19th century, but records are thin and unreliable, and certainly no paper has been published without interruption for as long as the French *Annales des Mines*. For some 10 years or so before the Russian revolution, *Mining Journal*'s Russian correspondent was Captain Eugene de Hautpick, a member of the Czar's General Staff and a prominent mining geologist. He also happened to be a close friend of Edward Baliol Scott and was Ursel Baliol Scott's godfather. He made frequent visits to Britain and after the revolution moved to Australia, where he continued his geological work. But he was a marked man, and in the early 1920s the long arm of the Bolsheviks reached out to him in Australia and he was eliminated. There cannot be many other cases of leading mining men suffering political assassination.

George O. Argall jnr, *World Mining*'s senior editor, has himself visited more than 60 countries in his 28 years with the paper, and in 1969 his report on mining in Japan won the Jesse H. Neal Editorial Achievement Award. He comes from an old family of mining engineers, originating in Cornwall, and of which one of the most distinguished members was his grandfather Philip, who gave young Rickard his first managerial job in California nearly 100 years ago (Chapter 11) and was a prolific writer on mining matters—including for Rickard's *Mining and Scientific Press*. George Argall's practical mining experience was obtained with underground and open pit operations in the western United States after he had graduated from the Colorado School of Mines, a career interrupted by service with the U.S. Navy's Seabees in the Pacific during the Second World War. Despite its wide international coverage and its overseas offices, *World Mining*, with its roots firmly planted in the largest metal-producing country, cannot avoid being influenced to some extent by events and opinions in that country. The ideal position from which to run an international publication is obviously one of neutrality. *Mining Journal* and its sister *Mining Magazine*—both now edited by John Spooner, who has been with the group for 12 years—are fortunate to be in such a position, for Britain today has little metalliferous mining, and its coal production, though substantial, is a matter of purely domestic concern and seldom occupies much space in either publication. Moreover, London is—as it has been for centuries—one of the world's great centres of communications, and it is the focal point of international metal marketing.

The expansion of the *Mining Journal* group in the international field over the past 10 years or so has been helped by these two factors, coupled with the effects of a change in the company's ownership. Anxious to secure the long-term independence of *Mining Journal*, Ursel Baliol Scott arranged to sell the company to Michael West, who was then editorial director. West, a Royal School of Mines graduate, who worked on the Zambian copperbelt as an engineer and chief of study, joined the paper at the end of 1960 to become its financial editor. In 1966, he became chairman and owner, and thus ended an era in which for 65 years *Mining Journal* had been in the hands of the Baliol Scott family, though a working partnership continued between West and Ursel Baliol Scott until the latter's retirement in 1978.

Perhaps the most significant feature of the change was that for the first time in its long history the Journal was now owned and editorially directed by someone with practical experience of the industry which it served. The new owner naturally had a keen appreciation of the problems of mining, having been personally confronted with many of them in a career which had taken him underground as an engineer and to a managerial position on surface. Ursel Baliol Scott has said that he and his staff tried to write from the standpoint of people in the industry and that, under his direction, the company regarded itself as "the publishing end of mining rather than the mining end of publishing". Certainly he went a long way towards achieving these aims, and during his editorship it became apparent that the industry was increasingly using the paper as a vehicle of communication within itself. However, when Michael West took over as editor, the change to professionalism—professionalism in mining engineering, though not necessarily in editorial know-how—was complete. Whilst this was a new feature in London-based mining journalism, in other countries it had already become the rule rather than the exception; in the United States, for example, professional mining men have occupied the *E/MJ* editorial chair throughout most of its hundred years of history.

But for *Mining Journal* it was a fundamental change, a change that has been further underlined during West's ownership, and is now reflected in its current policy to run its publications with its senior, editorial team exclusively composed of people with mining experience. As if to ensure that its readers are aware of this, each name in the credits box is followed by an impressive assembly of initials denoting academic achievement in one or other of the mining schools or universities and/or membership of some eminent professional institution.

This liking for what has been delightfully referred to in the United States as "caudal appendages", goes back more than 50 years in British mining journalism, certainly to 1924 when *Mining Magazine* had its leg gently pulled by an *E/MJ* leader writer who asked if so lengthy a justification of the standing of the paper's contributors was necessary, even assuming a general appreciation of such initials among English-speaking peoples. "We tremble at the possibility of error in deciphering the meteoric tails used by some of our British colleagues", wrote *E/MJ*. "Sooner or later we are sure that a grave error will be made in the print shop, if not in the editorial rooms. If the name of an author who contributes to current technical journalism does not carry sufficient weight, no amount of trimmings will save the situation. The test of greatness is the ease with which the name can be remembered, and titles and decorations can be forgotten, by posterity. And the most modest and effective manner in which such greatness may be anticipated is by the adoption of simplicity in matters of personal advertisement." How true! But what was then a curiously British trait has since been adopted by other mining papers—but not by *E/MJ*, of course!

The change of character in editorial direction—comparatively recent in the case of *Mining Journal*, but much earlier in evidence in many other publishing houses—has resulted in a more sensitive and appreciative treatment of mining news compared with the old days of editorship by non-professionals of the industry. It has also sharpened selectivity, and this has been particularly important with the increasing volume of available material as communications within the industry have developed. Even 50 years ago, one editor, eyeing an overflowing in-tray, petulantly scribbled: "Today it is not a question of lack of technical literature, but of its excess which makes it ever-increasingly difficult, if not impossible, for a mining engineer to keep abreast of anything but a very narrow section in his field."

Since then, the volume has continued to swell, and today editorial offices receive thousands of pieces of mail each year, including new books and periodicals; geological and mining reports; manufacturers' catalogues and announcements of new products; market reports; convention and symposium papers; patent reviews; annual and quarterly reports of mining companies all over the world, together with all the other company publications which their zealous public relations departments produce; articles submitted for publication by members of the profession; and, of course, an overwhelming amount of small print from government departments and international bodies. Buried in this stack are hundreds of significant items that can have a profound effect on the technology and future of the mining industry. The material in this cumulative pile must therefore be treated with great respect. Reader interests vary from the specialist who wants to know every tiny detail, to the broader range of mining people anxious to ensure that they do not miss anything important and who would prefer to have several articles in crisp, condensed form, rather than see a subject treated more expansively in a single feature. This is where the expertise of the mining professional comes in. In the quality journals, where there is always pressure on editorial space, he is obviously the best judge of what will appeal most to his colleagues in an industry which, despite its modern technical and geographical spread, often appears to be as tight a community as the guilds and chapters of Agricola's time.

Unfortunately, he is not always so well qualified to put his decisions into effect. There would appear to be few mining engineers around today with such a flair for writing and editing as, say, Raymond and Rickard, or even Rothwell; this is not surprising as the profession of journalism can no more be easily "picked up" than engineering or metallurgy. Some papers, among them *South African Mining and Engineering Journal*, accept that while qualified engineers and metallurgists are an essential part of an editorial team to maintain the quality and integrity of the main news and features, professional journalists are also necessary to ensure an acceptable level of writing and presentation. The economics of modern publishing do not, of course, always permit this, but it must surely be in the interests of publishers to ensure more thorough "on the job" training of editorial recruits from the mining industry than is available in most offices today. For a professional journalist, it is also saddening to reflect that not many years have passed since editorship represented the zenith of every young writer's ambition and, if this were ever achieved, it usually did not come until fairly late in life, after considerable experience in every aspect of the craft. Today, it is possible in some parts of the world for a man with a few technical articles to his credit to come straight off shift to join a team of "editors" of a mining publication. It must, of course, be emphasized that this downgrading of a title which used to represent a combination of journalistic experience, discipline and responsibility, is not peculiar to the mining world: it is, alas, discernible in much technical publishing today.

With the growth of a more technically-orientated type of journalism in mining periodicals, editorial crusading of the robust nature practised by people such as Henry English, B.T.A. Bell and the Baliol Scotts has now almost entirely disappeared. A study of the pages of contemporary periodicals, compared with those of 75, 100, or 140 years ago, might well give an impression that the spirit has gone out of the enterprise. At first sight, today's papers appear to have lost their virility; editorials are without punch, and time-honoured words and phrases have lost their meaning in a rash of technical jargon which sometimes suggests that it has spewed from a computer which has been improperly programmed. For this, there can be no excuse.

But the main reason why the thunderers are silent is that most of the major causes for which they fought have now been won, or lost, or finally abandoned. For example, the industry's morals do not now have to be so closely watched as in the days when fraud was rampant and shareholders' money was often used to fill the pockets of a few smart promoters. This is not to suggest that all mining companies today behave impeccably, but the chances of them offending against either law or ethics have become more remote—partly because the vigorous campaigning of the pioneer mining papers has contributed to the build-up of machinery which inhibits malpractice. Similarly, most of the world's big mines are as safe as engineering skill can make them; the health of those who work in them is protected to the limits of medical knowledge and research; and the education and training of potential engineers, metallurgists and geologists is, by and large, adequate for present requirements, even if the long-term future may give cause for some concern.

Earlier in this book (Chapter 17) it was suggested that every paper becomes the creature of the man who writes it, and that every business becomes the style of the man who controls it. With one or two exceptions—of which the *Mining Journal* group and *Skillings' Mining Review* are examples—most mining papers today are no longer editor-owned or editor-published. The majority are in the hands of large companies which produce other publications—groups such as McGraw-Hill, Thomson's and Miller Freeman. Consequently, the fortunes and policies of those papers in the group which are devoted to mining interests cannot be considered in isolation; they are part of the group, and when the board of directors meets it naturally takes first into account the contribution that these papers make to the group, rather than to the industry which they serve.

Vigorous editorial policies are usually initiated by owners and owner-editors, not by "independent" editors, and in these days owners tend to be much more concerned with advertising revenue and net profits than with pursuing moral or educative causes. This largely accounts for the fact that the old-style leading article, in which the positive personal views of the editor were expressed, has now been discontinued in most papers. It has been replaced in some periodicals by the sort of "Notes and News" column in which the opinions of others are quoted and the reader is often left wandering in a maze of conflicting argument when what he really needs is the guidance of one who is in a position to give an authoritative and independent view. This, after all, is what journalism is, or should be, all about. Some mining papers today do not even have an opinion column, either for themselves or for their readers, and stoically maintain a Vicar of Bray attitude to the industry.

Canada's *Northern Miner*, which is still very much a family concern, having been built up by the father and uncles of the current president, Richard C. Pearce, has recently been criticized for being old-fashioned; it was said that its approach to mining was that of a boom town chamber of commerce. Nothing can be further from the truth. The paper, which has retained its enterprise and virility over 60 years, has always been in a rather different category from its contemporaries. "We have never been a technical magazine in the same sense as *Mining Journal* or *World Mining*", says John Carrington, its editor from 1949 to 1978. "When people say they don't read the *Miner* because it is too technical, we feel that we are not doing our job. Of course, we carry a number of technical articles that are of interest only to operating personnel; that is because we try to have a balanced mix—something of interest for everyone.

"In our reporting of mining news we are presenting a technical subject in terms that are understandable to the layman. At the same time, we don't want to make it so simplified that the technical reader finds it unpalatable. We think we have been reasonably successful in this, and accordingly we have the support of the technical people in the industry as well as being read by the general public who have a mining interest. In matters of trade, politics, taxation and other general topics, we hope that it can be understood by any group of people."

For other papers also the old spirit is still alive, though it finds its expression in very different ways from the torch-bearing of the patriarchs. Rothwell once said that the prime function of *E/MJ* was to hold a mirror in which the industry could see a reflection of its development, but both he and his successors down the years have done a great deal more for the industry than that.

At *Mining Journal*, Michael West likes to feel that if the wrongs against which the old-time editors campaigned still existed, then he and his contemporaries would be acting in much the same way as they did. He also believes that if the pioneers were to return they would be affected by the same inhibitions which make things difficult for the editors of today.

Paradoxically, some of the most important of these inhibitions are linked with the mining professionalism which a place in the modern editorial team demands. For the past seven or eight years, the Institution of Mining and Metallurgy in London has recognized that a geologist, mining engineer, or metallurgist who edits a magazine that is intimately concerned with the industry, is in fact continuing to practice his profession; provided that he has previously "qualified" within the industry he may be elected to corporate status in the Institution and, of course, has the right to put its initials after his name. Such association is said to add considerably to the editor's, and his paper's credibility, and certainly it can give him a knowledge and appreciation of certain matters that could be denied him were he not a member. On the other hand, it could inhibit him from publishing views which may not conform with those of the Institution. Immediately, therefore, he is faced with a conflict of divided loyalties—a conflict which seldom arises in newspaper journalism in which the first professional responsibility of any individual is accepted as being directly to his paper.

However, despite such constraints, much of the mining Press continues to make a valuable contribution to thought and opinion in the industry but, as already indicated, it does not do so in the same rumbustious manner as in the past. When 95 miners died in the Haswell disaster in 1844, Henry English rushed to Buckingham Palace to petition Queen Victoria to use her royal prerogative to set up a Commission of Inquiry into the appalling accident rate in British mines. When, on September 25, 1970, 89 underground workers lost their lives as a sea of mud poured through a sinkhole at the Mufulira copper mine in Zambia, the *Mining Journal* group immediately reported the accident in detail in its airmail weekly, but waited nearly six months, until all possible evidence had been collected and published, before it offered comment. Then, in a carefully considered article in *Mining Magazine*, it emphasized that whilst engineering investigations must attempt to establish what went wrong, it was even more imperative to examine the human attitudes associated with the event, pointing out that familiarity with a situation often increases its potential dangers. The article went on to call for the introduction of a system under which mining and metallurgical operations would be visited periodically by "competent strangers", independent technicians and technologists, who would report direct to the chairman. In this way the chances of a repetition of the Mufulira tragedy might be substantially reduced.

Mining Journal has continued to highlight all aspects of mine safety and to campaign for the acceptance of the social audit concept in the industry. One day it may succeed. Whether it does or not, it will never be possible to compare the efforts of the paper's first editor with those of its present editorial director, and to say which have been the more effective.

Similarly, in the matter of technical education and training, *Mining Journal* today is as concerned as it was in Victorian times. In the 1960s it initiated a "conference by correspondence" which laid the foundations for a major investigation into the long-term manpower requirements of the industry, known as the "Brighton Conference". Out of this, the Mineral Industry Manpower and Careers Unit was born, funded in the main by London-based mining finance houses. The Unit's chief concern is to prepare forecasts of the industry's requirements in all professional staff categories and to take such measures as it can to ensure that they will be met. Whilst the Unit is not a recruiting agency, it is currently encouraging more students in Britain to study for mining degrees. It does this through a network of links with some 300 schools in what it calls its Phoenix Programme, which is essentially an information service about the arts and science of minerals, and the functions and responsibilities of the industry's decision-makers. Phoenix has worked well enough for similar programmes to have been started in South Africa, but its main constraint is the limited number of institutions offering courses in the disciplines of mining and metallurgy in the United Kingdom. Incidentally, while *Mining Journal* cannot claim to have been directly concerned in helping to launch any new professional bodies since its part in founding the IMM some 80 years ago, it made its offices available, and provided secretarial assistance, for a series of meetings which resulted in the establishment in 1974 of the Institute of Professional Geologists.

One of the most significant contributions which elements of the specialist Press have made to the industry during the past 15 years or so has been the sponsorship of meetings and conferences which have not only given the major problems of the industry public airing, but, perhaps more importantly, have brought together representatives of many different branches of the industry, often for the first time. In particular, such conferences have encouraged a much closer relationship between producers on the one hand, and market-men and consumers on the other.

Pioneer work in this field by *American Metal Market* and its enterprising former chairman, Archer Trench, has already been mentioned (Chapter 17), but a number of other publishing houses have since become quite heavily involved in the conference business. To some extent, this is an exercise in self-interest, and in a few cases the commercial rewards are considerable, but by and large the main motive of the sponsors has been to provide a facility rather than to make a profit.

Most conferences are well thought out and well run, but *E/MJ* undoubtedly voiced the opinion of a large number of convention-goers when it suggested that all papers submitted to such gatherings should be pre-printed, complete with illustrations, so that the audience is not obliged to squint at unrecognizable slides and take notes. "If a paper is not worth pre-printing," said *E/MJ*, "it is not worth giving—that was the opinion of mining engineers and consultants who attended a recent regional convention, at considerable personal or company expense, only to find that they could not have access to the information which was promised on the program.

"Conventions and symposia should also be planned to give those who attend an opportunity to visit and discuss ideas at leisure with fellow members of the profession."

The importance of free-flowing communications becomes evident when it is realized that more than 250 major ideas were developed or introduced in international mining in 12 years following the Second World War. This in itself is a justification of the growth of any system that encourages the exchange of information and opinion. The only danger is that in certain aspects the industry would appear to be reaching a saturation point in conferences, and professional men with limited time will have to become more and more selective when they examine the countless prospectuses that arrive through the mail. On the other hand, it is probably true to say that the market has hardly been scratched for those conferences which are designed to provide a forum for people in the smaller and more specialized sectors of the industry who have a natural desire to come together and debate their problems.

For the sponsors, it must also be recognized that such conferences provide a guarantee against any shortage of copy for quite a few issues.

Finally, in considering the ways in which the mining Press can serve the industry today, the participation of certain of its members in the councils of international bodies, including the various agencies of the United Nations, must not be overlooked. One editor orders his life and his work to enable him to spend a certain amount of his time at the U.N. He deeply believes that the minerals industry has an important role to play in its deliberations and policies, and it is apparent that those who sponsor him consider that they could do no better than to have an ambassador who has been trained in the disciplines of the industry, has observed the world in mining terms and has the international perspective and independence that can best be provided by a background in technical journalism.

*

The origins of mining and metallurgy go back deep in time. Somewhere in the misty dawn of civilization, man hacked at the earth's crust with a stone hammer and chipped away a piece of mineralized rock. He crushed it into powder, mixed it with fluxes and roasted it in a charcoal fire in a hole in the ground, thus evolving a technology which remains the basic principle of metal smelting today.

Some say that the first metallurgists worked on the banks of the Tigris or Euphrates, in a land east of Babylon which was once called Elam. Or perhaps they lived amidst the gaunt majesty of Sinai, where the remains of rough smelters have been found, dating back 6,000 years; perhaps in the Far East, or even in Western Europe. Timna, a great rocky enclave in the red sandstone mountains of the Negev desert in Israel, was an important copper producer for the Egyptians in the 14th–12th centuries BC, even if recent archaeological discoveries have exploded the myth that it was also the site of King Solomon's mines. The mines of Laurium, in southern Attica, were already ancient in the days of Xenophon, and their silver played an important part in the economy of Athens. Of equal significance was the gold found in the gravel of the river Strymon in Thrace, and the silver in the limestone of Mount Pangaeus that filled the treasury to finance the wars of Philip of Macedonia and his great son Alexander. The Spanish mines were an important element in the economy of ancient Rome; Pliny wrote that they yielded Hannibal 300 lb. of silver a day to provide the backing for his famous military adventures.

How and precisely when the industry began we shall never know for sure. But as soon as men learned to write, they set down the record, at first in hieroglyphics on tablets, on parchment and on paper—in scrolls, in books, in journals and in newspapers. Their writings have ranged from those of Aristotle and Theophrastus, to Strabo and Pliny; from Albertus Magnus to Agricola; from Werner to Hoover; from

Henry English, Raymond and Rothwell, to the Baliol Scotts and the editors and writers of mining literature today. Mining is indebted to them all, not only for its history, but for many other contributions offered from a vantage point a little distance from the problems which sometimes excite and disturb it, but never far from the beating of its heart.

So much for the past. What of the future?

Undoubtedly, mining's special media will continue to provide the means whereby the industry can see the reflection of its development. It will hold its mirror to operations and geological exploration in every part of the world; it will record technology and methodology undreamed of a century ago; it will reflect progress towards the mining of minerals from new sources, including from the bed of the ocean, and perhaps even from the planets revolving in outer space.

More immediately, it will seek to extend its services and its readership in the developing countries which together currently produce something like one-third of the world's total mineral supplies. As has been seen, great changes have already taken place over the past 20 years in the ownership of ore reserves and mines in the Third World. Inevitably, the involvement of the public sector in world mining will continue to increase; the big problem is to ensure that it does so peacefully and that the policies and outlook of the new owners will mature with the responsibilities that they assume.

In this transitional period the media has an important part to play, not necessarily in attempting to teach either side its responsibilities— for that would be impertinent—but, by its dissemination of news and opinion, in helping to bring about a deeper appreciation of each party's aspirations, fears and policies. If nothing else, today's mining Press can provide a means of communication that is in tune with the industry and can offer a platform for discussion without the necessity of rushing every issue to UNCTAD, or similarly well-meaning, though impersonal bodies.

Four hundred years ago, Agricola lamented that the popular concept of mining was "a business requiring not so much skill as labour." *De Re Metallica* went a long way to disabuse that impression, and in the years that have followed, other writers, other books and other journals have done much to put the industry into its proper perspective in the context of developing civilization. Today, it is recognized as one of the world's largest undertakings in the interests of mankind, a link between nations, its products supplying the lifeblood of our industrial society. In the transition, its chroniclers have played a major part; in the solution of the problems that lie ahead, their role will be no less important.

BIBLIOGRAPHY

Agricola, Georgius, *De Re Metallica* (1556), Transl. Hoover, H.C. and L.W. (1912)

Ballantyne, Rev. James, *Homes and Homesteads of Victoria*

Battiscombe, Georgina, *Shaftesbury, a Biography of the Seventh Earl, 1801–85*

Cartwright, A.P., *The Gold Miners*

Dibner, Bern, *Agricola on Metals* (1958)

du Maurier, Daphne, *Vanishing Cornwall*

Hodder, Edwin, *The Life of a Century* (1901)

Hoover, H.C., *The Memoirs of Herbert Hoover, 1874–1920, Years of Adventure*

Koch, Dr.-Ing. Manfred, *Geschichte und Entwicklung des bergmännischen Schrifttums*

Lyman, George D., *The Saga of the Comstock Lode* (1934)

Prain, Sir Ronald L., *Selected Papers* (4 volumes, 1953–67)

Reeks, Margaret, *Royal School of Mines—Register of Old Pupils and History of the School* (1920)

Rickard, T.A., *Man and Metals* (1932); *Retrospect* (1937)

Rothwell, R.P., *The Mineral Industry, its Statistics, Technology and Trade* (1892–1942)

Skinner, *Walter R. Skinner's International Mining Year Book*

Sloane, N. and Lucille L., *A Pictorial History of American Mining*

Stevens, Horace J., *The Copper Handbook*, Vol.2 (1901), Vol.3 (1902)

Weed, Walter Harvey, *The Copper Mines of the World* (1907)

Transactions of:
American Institute of Mining, Metallurgical and Petroleum Engineers; Canadian Institute of Mining and Metallurgy; Institution of Mining and Metallurgy

Publications of:
American Mining Congress; Australian Mining Industry Council; Conseil Intergouvernemental des Pays Exportateurs de Cuivre; Copper Development Association; Mining Association of Canada; U.S. Bureau of Mines; World Bureau of Metal Statistics

Mining Periodicals:
American Gold News; American Metal Market; American Mining Congress Journal; Annales des Mines; Arizona Mining Journal (1917–1946); *Australian Mining; Australian Mining Standard* (1889–1961); *Australian National Miner;*

Californian Mining Journal; Canadian Mining Journal; CIM Journal, Canadian Institute of Mining and Metallurgy; *Coal, Gold and Base Minerals of Southern Africa; Daily Metal Reporter,* New York (1911–1961); *Denver Mining Record; Engineering and Mining Journal; IMM Bulletin,* Institution of Mining and Metallurgy, London; *Indian Mining and Engineering Journal; Iron Age;*

Metal Bulletin; Metall, Berlin; *The Metallurgist,* Institution of Metallurgists, London; *Metals Week; Mine and Quarry; Mining Engineer,* Institution of Mining Engineers, London; *Mining Engineering,* American Institute of Mining, Metallurgical and Petroleum Engineers, New York; *Mining Journal; Mining Magazine; Mining and Scientific Press,* San Francisco (1860–1922); *Mining World,* London (1871–1968);

Northern Miner, Toronto; *Optima,* Johannesburg; *Pay Dirt,* Bisbee, Arizona; *Queensland Government Mining Journal; Skillings' Mining Review,* Duluth, Minnesota; *South African Mining and Engineering Journal; Western Miner,* Vancouver; *World Mining*

Newspapers:
Financial Times; New York Times; The Times; Rand Daily Mail; Wall Street Journal

Index

International Exhibition of Mining and Metallurgy, 1890, 67–9, 83, 106, 122
International Iron and Steel Institute, 215
International Mining Convention, Montreal, Canada, 1893, 87–8
Inter-Provincial Conference of Mining Engineers, Canada, 1897, 88–9
Investment, in mining industry, 251–2, 262
Investors Review, London, 241, 243
Iron Age, The, USA, 196–201, 203, 211, 220, 270
  price reporting, 196–7, 200–1
  metallurgical coverage, 198–9
  production figures, 200–1
Iron and Coal Trades Review, USA, 211
Iron and Steel Institute, 101
Iron and Steel Works of the World, 223
Ironmonger, The, London, 205, 220–1
Isaacs, Sir Henry, 68

Jameson Raid, 72–3
Jeans, J.S., 62
Joachimstal, mining town, Germany, 2
Joel, Solly, 115
Johannesburg Stock Exchange Official Gazette, 116–7
Johnston, William J., *E/MJ*, 138, 140
Jones, Roderick, Reuters, Johannesburg, 236
Journal Anglais (1775), 16
Journal des Mines de la République (1794), 17–21, see also *Annales des Mines*
Journal des Savants (1655), 16
Journeys of Observation, T.A. Rickard, 146
Julian, E.W., manager, Mining Publications Ltd., 186

Kalgoorlie, Western Australia, 108, 137
Kazakhskaya, USSR, mining in, 271
Kelly, Cornelius, Anaconda company, 228
Kemp, James F., 62, 144
Kimberley, South Africa, 65
Kimberley, Western Australia, 108
Kinsman, Victor, 112
Kirkland Lake, mining area, Canada, 156

Kirkpatrick, Professor S.F., 91
Klondike, gold rush, 73, 78
Knoerr, Alvin W., *E/MJ*, 177–81, 232–3
  uranium campaign, 178–81
  editor, 177–81
  market reporting, 206–7
  centenary issue, 177
  on Dr. Zimmerman, 232–3
Knutsford, Lord, 68
Köhler and Hoffmann, publishers, 16
Kolar, gold field, Mysore, India, 82
Kruger, Paul, President of South African Republic, 116
Kunz, George, 62
Küstel, Guido, 49

La Feuille Nécessaire (1759), 16
La Restitution de Pluton (1640), Martine de Bertereau, 14
Ladies' Forum Club, London, 182
Ladies' Home Journal, San Francisco, 46
Lafayette College, USA, 97, 101
Lake copper, 53
Langeloth, J., 62
Langer, Norbert, *American Metal Market*, 212, 213
Laurium, ancient mine, Greece, 282
Lavoisier, Antoine, chemist, 15, 20
"Law of the Apex", 99–100
Lawton, William, explorer, Australia, 121
Leadville, Colorado, USA, 59, 134, 136
Leblanc, Nicolas, chemist, 20
Lehigh University, USA, 101
Leipzig Magazine (1781), 16
Leipzig, University of, 2, 96
Lempe, Johann Friedich, publisher, 16
Levin, Bernard, Fleet Street columnist, 255–6
Lincoln, Abraham, President, United States, 42, 236
Lionaise, Charles, 76
Lipsett, Charles H., Atlas Publishing Company, New York, 212
Logan, Sir William E., 89
London Metal Exchange, 53, 201–2, 239
  annual dinner, 216
London Stock Exchange, 34
Longwell, Alexander, 90
Luther, Martin, 2

INSTITUTIONS AND ASSOCIATIONS

PUBLICATIONS

Books and Handbooks on mining, minerals etc.

General Literature